Parallel Processing in Structural Engineering

Parallel Processing in Structural Engineering

Professor Hojjat Adeli

The Ohio State University, Columbus, Ohio

and

Dr Osama Kamal

Zagazig University, Cairo, Egypt

ELSEVIER APPLIED SCIENCE
LONDON and NEW YORK

ELSEVIER SCIENCE PUBLISHERS LTD
Crown House, Linton Road, Barking, Essex IG11 8JU, England

WITH 29 TABLES AND 113 ILLUSTRATIONS

© 1993 ELSEVIER SCIENCE PUBLISHERS LTD

British Library Cataloguing in Publication Data

Adeli, Hojjat
 Parallel Processing in Structural
 Engineering
 I. Title II. Kamal, Osama
 624.10285

ISBN 1-85861-003-6

Library of Congress Cataloging-in-Publication Data

Adeli, Hojjat, 1950–
 Parallel processing in structural engineering/Hojjat Adeli and Osama Kamal.
  p.   cm.
 Includes bibliographical references and index.
 ISBN 1-85861-003-6
 1. Structural analysis (Engineering)—Data processing.
 2. Parallel processing (Electronic computers)   I. Kamal, Osama.
 II. Title.
 TA647.A34 1993
 624.1'71'0285435—dc20   92-42747
                                                            CIP

No responsibility is assumed by the Publisher for any injury and/or damage to persons or property as a matter of products liability, negligence or otherwise, or from any use or operation of any methods, products, instructions or ideas contained in the material herein.

Special regulations for readers in the USA

This publication has been registered with the Copyright Clearance Center Inc. (CCC), Salem, Massachusetts. Information can be obtained from the CCC about conditions under which photocopies of parts of this publication may be made in the USA. All other copyright questions, including photocopying outside the USA, should be referred to the publisher.

All rights reserved. No part of this publication may be reproduced, stored in a retrieval system, or transmitted in any form or by any means, electronic, mechanical, photocopying, recording, or otherwise, without the prior written permission of the publisher.

Photoset and Printed in Northern Ireland at the Universities Press (Belfast) Ltd.

Preface

This book summarizes our research on the development of parallel algorithms for the analysis and optimization of structures performed during the past few years at The Ohio State University and published in the journals of *Microcomputers in Civil Engineering, Computers and Structures, ASCE Journal of Aerospace Engineering,* and *International Journal of Mini and Microcomputers,* and presented at several conferences.

Parallel processing is currently an area of intensive research. We shall see a lot more exciting research in the coming years. Concurrently, we should see more widespread applications of parallel processing in structural engineering. As we demonstrate in this book, parallel processing is particularly attractive for analysis and optimization of large structures. We hope that this first book on the subject will lead the way and encourage further research and development in this rapidly growing area of high technology.

HOJJAT ADELI and OSAMA KAMAL
January 1992

Acknowledgement

Part of the material used in this book is based on the authors' own research papers that have appeared in journals published by Elsevier, Pergamon Press and the American Society of Civil Engineers, and in conference proceedings published by the Institute of Electrical and Electronics Engineers, as included in the references at the end of the book, and are reprinted with their permission.

*Dedicated
to
Our Wives*

Contents

Preface		v
Acknowledgement		v
Chapter I	**Introduction**	1
Chapter II	**Basic Concepts**	3
2.1	Introduction	3
2.2	Encore Multimax	3
2.3	Programming Language	5
2.4	Structural Analysis Problem	5
2.5	Threads	7
2.6	Racing Condition	8
2.7	Synchronization	9
2.8	Mapping	10
2.9	Parallel C Program PASTRANC	12
2.10	Applications	14
2.11	Summary and Conclusions	30
Chapter III	**Automatic Partitioning of Framed Structures for Concurrent Processing**	32
3.1	Introduction	32
3.2	Basic Concepts and Definitions	36
3.3	Pre-partitioning	42
3.4	Initial Partitioning	44
3.5	Intermediate Partitioning	49
3.6	Final Partitioning	52
3.7	Post-partitioning	57
3.8	Applications	60
3.9	Summary and Conclusions	68

Chapter IV Concurrent Analysis of Structures 74
4.1 Introduction 74
4.2 Equilibrium Equations 76
4.3 Assembling the Structure Stiffness Matrix 78
4.4 Load Vector 82
4.5 Boundary Conditions 83
4.6 Static Condensation 83
4.7 Interface Linear Equations 88
4.8 Non-interface Displacements 92
4.9 Forces and Stresses 92
4.10 Applications 92
4.11 Summary and Conclusions 107

Chapter V Concurrent Optimization of Structures . . . 111
5.1 Introduction 111
5.2 An Optimality Criterion Approach 113
5.3 Algorithms for Parallel Structural Optimization . . 117
5.4 Applications 122
5.5 Summary and Conclusions 138

Chapter VI Concurrent Optimization of Structures under Dynamic Loading 140
6.1 Optimization under Dynamic Loading 140
6.2 Assembling the Structure Mass Matrix 143
6.3 Solution of the Eigenproblem 148
6.4 Temporal Solution 149
6.5 Dynamic Response 152
6.6 Applications 152
6.7 Summary and Conclusions 171

Chapter VII Conclusions 175
7.1 Summary and Conclusions 175
7.2 Future Research 177

Appendix: **Sample Code** 179
References 180
Index 184

Chapter I

Introduction

Parallel processing provides an opportunity to improve computational efficiency by orders of magnitude. By combining 100 microprocessors each with performance of 100 MIPS (million instructions per second), it is possible to gain supercomputer performance at a fraction of the cost of a supercomputer. The research challenge is to reformulate the problem, develop parallel algorithms, and devise new computational stratagems in order to fully utilize the capabilities of parallel machines (Adeli & Vishnubhotla [1987]).

Various types of multiprocessor computers have recently become commercially available. Surveys of these machines can be found in Adeli & Vishnubhotla [1987, 1992] and Dongarra & Duff [1992]. A number of researchers have recently presented algorithms for concurrent processing of structures. Chien & Sun [1989] and Lou & Friedman [1989] have presented algorithms for concurrent analysis of structures modeled by finite elements. Research attempts on the concurrent dynamic analysis of structures have been presented by Lo & Phillipe [1986], Farhat & Wilson [1987], Bostic & Fulton [1988], and Noor & Peters [1989]. Little work has been reported in the area of concurrent optimization of structures. Sikiotis & Saouma [1987] outlined a framework for parallel structural optimization on a network of engineering workstations. They suggested distributing gradient computations at each iteration among several computer workstations. Svensson [1987] proposed a substructuring approach to optimum structural design based on an active design variable strategy and outlined how this strategy can be extended to run on multiprocessor computers.

The aforementioned research has been performed on a wide variety of multiprocessor computers, including the FLEX/32, Intel iPSC hypercube, Sequent S series, Encore Multimax, Butterfly Plus, Alliant FX/8, Cray X-MP, and Cray Y-MP. Descriptions of these computers can be found in Adeli & Vishnubhotla [1987, 1992] and Dongarra & Duff [1992].

In this book, we present efficient parallel algorithms for the analysis and optimization of structures. We reformulate the structural analysis and optimization problems, rearrange the computations, and devise new stratagems utilizing the concurrent processing capabilities of new multiprocessor computers. The focus of the presentation is on framed structures. We also study the various factors affecting the concurrent performance of the algorithms, such as the overhead time spent in achieving parallelism, synchronization, and context switching.

In Chapter II, we introduce the computing environment and basic parallel processing concepts, and investigate the effects of various factors on the efficiency of concurrent algorithms for analysis of structures. In Chapter III, we present a general substructuring (partitioning) algorithm for parallel processing of framed structures. Chapter IV deals with the formulation and algorithms for concurrent analysis of framed structures. A new optimization formulation and parallel algorithms for optimization of structures are presented in Chapter V. Chapter VI discusses the concurrent optimization of structures under time-dependent dynamic loading. Finally, Chapter VII draws conclusions and ends with final comments on extension of this work.

Chapter II

Basic Concepts

2.1 INTRODUCTION

In this chapter, we first review the basic concepts and tools for concurrent processing and introduce the notion of cheap concurrency and the concept of thread. Subsequently, we investigate the effects of various factors on the efficiency of concurrent algorithms for the analysis of structures.

Section 2.2 introduces the computing environment: the Encore Multimax shared-memory multiprocessor computer. Section 2.3 describes some of the features of the C programming language used in this work. Section 2.4 summarizes the structural analysis problem, main equations, and the serial sequence of solution. The concept of threads is introduced in Section 2.5. Following this, Sections 2.6–2.8 describe issues involved in concurrent processing, such as racing condition, synchronization, and mapping. The program PASTRANC (PArallel STRuctural ANalysis in C) is described in Section 2.9. Applications and comparative results are given in Section 2.10. Finally, conclusions are presented in Section 2.11.

2.2 ENCORE MULTIMAX

Encore Multimax (Encore [1985]) is a multiprocessor computer with an upper limit of 20 processors. Figure 1 outlines the architecture of an Encore Multimax. The system is configured in terms of dual processor cards; each card has two processors and a cache memory (i.e. the modular growth in the number of processors is in increments of two).

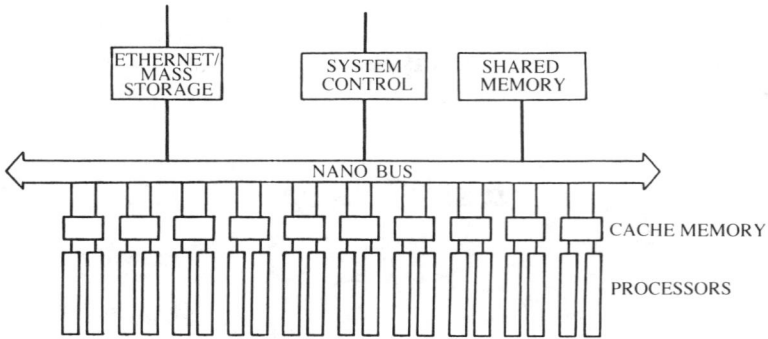

Fig. 1. Schematic architecture of Encore Multimax.

Each processor is a National Semiconductor NS32032 32-bit processor operating at 10 MHz and capable of executing two million instructions per second (MIPS). Each two processors share a single 32 Kilobyte cache memory. Shared memory is scalable from 4 Megabytes to 64 Megabytes in increments of 4 Megabytes. Input/output (I/O) is scalable from 1·5 Megabytes/s to 15 Megabytes/s in increments of 1·5 Megabytes/s. The system uses a very fast bus, Nanobus, that can carry 64 bits of new information every 80 ns, even when previous requests are in progress. This results in a true data transfer bandwidth of 100 Megabytes/s. Thus, the Nanobus secures fast synchronization between processors and I/O devices, with negligible effect on the performance of other system activities.

Whenever either one of the two processors of a certain cache attempts to read from or write into a main (shared) memory location, the memory data will be automatically written into their 32 Kilobyte cache memory. Thereafter, any attempt by either of the two processors to access those main memory locations will be directed automatically to their corresponding locations in their cache. In addition, Encore Multimax devices keep the cache of a dual processor card updated with any changes (due to writes issued by any other processor or device) that might occur to a main (shared) memory location. In this sense, the cache memory of a processor services approximately 95% of a processor's requests for memory. The remaining memory requests are fulfilled through the main memory, thus reaching the bus. In general, Encore Multimax preserves a linear proportionality between system performance and number of processors as long as the bus is not saturated with main memory access requests. Encore Multimax supports UMAX 4.2, UMAX 4.3, UMAX V, and Mach operating systems.

2.3 PROGRAMMING LANGUAGE

In the early 1970s the C programming language was originated at Bell Laboratories by Dennis Ritchie (Kernighan & Ritchie [1988]). It was aimed to be the standard programming language for the UNIX operating system. However, the language became popular only in the late 1970s, when UNIX was released for commercial use. On one hand, C may be considered a higher-level programming language in the sense that it is hardware- or machine-independent. Also, it is intended to be independent of the operating system (Kernighan & Ritchie [1988]). On the other hand, C is a relatively low-level language in the sense that it helps the programmer deal with the computer at a relatively low level that is closer to the hardware. In general, C is a general-purpose structured programming language that provides the programmer with a lot of flexibility and power. This is due to the fact that C has a number of powerful features (e.g. pointers) that set it apart from other popular programming languages such as FORTRAN and BASIC. Consequently, some application programs that are extremely difficult to code in FORTRAN, BASIC, or even Pascal can be easily coded in C (Kochan [1988]).

2.4 STRUCTURAL ANALYSIS PROBLEM

Using the stiffness method, the linear structural analysis problem can be summarized as follows. Assuming that the structure is discretized into a finite number of elements, the equilibrium equations can be written as

$$\mathbf{Kr} = \mathbf{R} \qquad (1)$$

where **K** is the structure stiffness matrix of size $k \times k$, k being the number of generalized displacement degrees of freedom of the structure. The column vector **r**, of size $k \times 1$, represents the structure generalized displacements vector. The set of generalized external forces applied to the structure are represented by the load vector **R**, of size $k \times 1$.

The structure stiffness matrix **K** is set up by adding up the contribution of the stiffness matrices of individual elements. Similarly, the generalized load vector **R** is composed of load vectors for all the elements. Once this assembly process is completed, the set of simultaneous equations (1) can be solved for the structure generalized displacements vector **r**. Next, the set of element internal forces S_i in member i can be determined through the equation

$$S_i = K_i r_i \qquad (2)$$

where K_i is the ith element stiffness matrix and r_i is the generalized displacement vector for the ith element. Finally, the stress vector σ_i of the ith element can be determined from the following relationship:

$$\sigma_i = B_i S_i \qquad (3)$$

where B_i is the stress transformation matrix containing the cross-sectional properties for the ith member.

Based on the previous equations, the sequence of steps involved in the serial (sequential) solution of the structural analysis problem is shown in Fig. 2. First, the data for different sections, nodal points, and elements are read. Next, the elements stiffness matrices are set up and assembled into the structure (global) stiffness matrix. The fact that this global matrix is symmetric and banded is used for storing it in banded

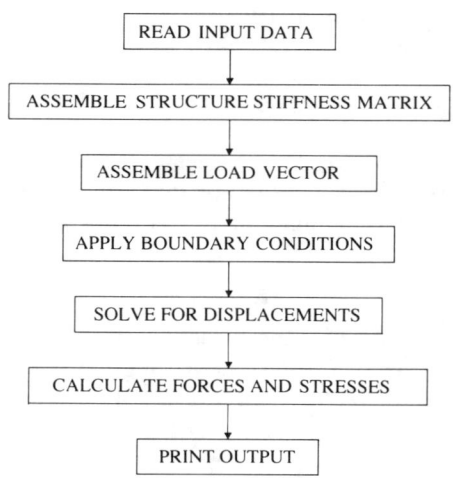

Fig. 2. Flow of a serial structural analysis program.

form. Then, the structure load vector is assembled from the load vectors of individual elements. The boundary conditions are applied, thus reducing the size of the set of simultaneous linear equations (1) to be solved. Once the structure generalized displacements vector **r** has been found, element forces and stresses can be evaluated through the use of eqns (2) and (3), respectively. Finally, the results are printed out.

2.5 THREADS

A thread is a unit of execution (agent executing instructions) that is independent of other similar units (threads), yet it can execute concurrently with them. The concept of threads was first developed by Thomas Deoppner [1987] at Brown University. The notion of a thread is quite different and independent of that of a processor. For example, there can be many threads running on one processor or concurrently on several processors. This results in a high level of abstraction to the programmer, who becomes separated from such details as the number of processors that are available. The programmer's concern needs to be limited to developing a certain number of threads. Deoppner had the goal of making concurrency a practical programming concept. Indeed, creating, scheduling, and synchronizing threads can be done at a relatively low overhead computational cost. Also, threads can be used for performing concurrent I/O. In short, threads provide a programmer with concurrency and abstraction at a low cost. This has led to the notion of cheap concurrency, which is also referred to as 'lightweight processes'.

In fact, Encore Multimax provides a programmer with even a higher level of abstraction. Once a programmer identifies the processes (threads, in this context), Encore will dynamically allocate the processors to the processes with currently highest priority. In addition, a programmer does not have any control of which process is assigned to which processor. In this sense, process–processor assignment is not influenced by the programmer.

Encore Parallel Threads (EPT) provide a set of constructs necessary for implementing the threads on an Encore Multimax (Encore [1988]). It is based on the Brown University Threads Package. It can be used with the C programming language under UMAX operating system. In general, EPT provides groups of constructs that support creation of threads, synchronization of threads through monitors or semaphores, creation of thread control blocks, raising exceptions, handling interrupts, and performing shared I/O. The constructs relevant to the work presented in this book are discussed later in detail.

Encore Parallel Threads (Encore [1988]) are used in this work for creating concurrency throughout the steps of assembling the structure stiffness matrix and evaluating the element forces and stresses. In addition, synchronization constructs are used to overcome the 'racing condition' problem that arises during the assembly process. A synchronization call issued to a thread suspends or resumes its execution on the processor it is assigned to. However, for clarity of discussion, we prefer to use the term 'synchronization of processors' rather than 'synchronization of threads'. Thus, discussion of synchronization in the rest of this book will be in reference to the processors.

2.6 RACING CONDITION

To describe a racing condition, consider the case of a processor writing into a shared memory location. When no other processor attempts to write into the same shared memory location simultaneously, Encore Multimax devices guarantee correct answers. What happens when a number of processors attempt to write into the same shared memory location, simultaneously? For example, suppose $x = 5$ is stored in a shared memory location, and one processor tries to execute the operation $x = x + 1$ and another one tries to perform $x = x + 2$, simultaneously. A sequential code would set the value of x to 8. But what happens if the two operations are executed concurrently? Since each operation is implemented within each processor in machine language as *load x, add* (1 *or* 2), *and store x,* one scenario is as follows:

1. First processor loads its copy of x; $x = 5$.
2. First processor performs $x = x + 1$: $x = 6$ (*but it is not stored yet*).
3. Second processor jumps in and starts by loading its copy of x: $x = 5$.
4. Second processor performs $x = x + 2$: $x = 7$ (*but it is not stored yet*).
5. First processor stores its current value of x into the shared memory: $x = 6$.
6. Second processor stores its current value of x into the shared memory: $x = 7$.

The outcome of this scenario is $x = 7$. Another scenario would result in $x = 6$. A third would result in $x = 8$, which is the correct solution. In short, the outcome is dependent on how different steps are interleaved or on the details of the 'race' between the processors. This raises the notion of a 'racing condition' or a 'critical section problem' (Peterson & Silberschatz [1985]).

In the structural analysis program, a racing condition arises during the assembly process. When two (or more) processors try to write, simultaneously, the stiffness matrices of their assigned elements into the same location of the shared memory for structure stiffness matrix, time-dependent errors and incorrect results may be the outcome. The remedy for such a racing condition is described in the following section.

2.7 SYNCHRONIZATION

A remedy for a racing condition or a critical section problem is that when one processor is executing in its critical section, all other processors are denied access to their critical sections. In other words, when a number of processors attempt to write into the same location of the shared memory simultaneously, all but one will have to wait in queue until this one processor finishes its engagement with this location. Then, the processor next in the queue gets in, and so on. The result is that the critical sections are executed by the processors mutually exclusive in time. This process is called 'synchronization'. A taxing consequence of synchronization is that it jeopardizes the linear speed-up sought for the system. In this chapter, two synchronization options are used and investigated: semaphores and monitors.

A semaphore (Dijkstra [1965]) is a synchronization tool that secures mutual exclusion in time in executing critical sections. A semaphore S is an integer value that can be accessed only through two standard atomic operations: $P(S)$ and $V(S)$. For example, the value of S can be 0 or 1. Also, when one processor modifies the S value, no other processor can simultaneously change this same value. When a processor is about to enter its critical section, it executes the $P(S)$ operation. This operation tests the S value for permission to access the critical section. Based on this value, the permission is granted or denied. When S is equal to 0, access is denied and the processor parks in queue at the top of its critical section. When S is equal to 1, the processor resumes its $P(S)$ operation and enters its critical section after setting the value of S to 0, thus denying all other processors access to their critical sections. Once the processor has finished executing within its critical section, it executes the $V(S)$ operation that resets the value of S to 1. This operation allows the processor to exit its critical section, and grants another parked processor permission to resume its $P(S)$ operation and enter its critical section.

In the semaphore solution, however, each processor must execute the P operation before entering the critical section and the V operation when exiting it. If this sequence is not observed (for instance, due to a programming error) by even one processor, time-dependent errors and incorrect results may occur. For example, if a processor executes the V operation before the P operation, more than one processor will be active in their critical sections simultaneously. Consequently, time-dependent errors may result. Other instances may lead to a deadlock situation. Therefore, the need appeared for some other language construct to protect against simple errors that might occur when using the semaphore solution for solving the critical section problem.

Hoare [1974] developed a language construct called 'monitor'. A monitor invokes P and S operations on semaphores in a manner that assures that only one processor is active within the monitor. This property is guaranteed by the monitor itself. In addition, a monitor provides support to variables of type 'condition'. These variables can be accessed only through two operations, namely, *wait* and *signal*. Whenever a processor invokes the operation *wait,* it is suspended until another processor invokes the operation *signal*. The operation *signal* causes one suspended processor to resume operation. Contrary to the V operation, the operation *signal* does not affect the state of the processor when no processor is suspended. Implementation of the *wait* and *signal* operations involves P and V operations on semaphores. In general, variables of type 'condition' allow programmers a good deal of flexibility to tailor their own synchronization patterns (Peterson & Silberschatz [1985]). In short, semaphores are used in cases of simple mutual exclusion while montiors are safer to use when exceptions are in existence. Comparison between the two synchronization schemes is included in Section 2.10.

2.8 MAPPING

As pointed out earlier, there can be many threads running on one processor or concurrently on several processors. Once the threads have been created and the number of processors specified, the scheduling (assigning) process, (i.e. which thread is assigned to which processor) is dynamically accomplished by the operating system. As a result, programmers are faced with a situation in which they cannot contribute to the process of binding the world of threads to the world of processors.

However, a programmer can contribute significantly to the efficiency of the solution process based on the strategy chosen for mapping the components of the concurrent application to the threads. For example, an application may have a large number (say, a thousand) components that can be processed concurrently. Hence, the programmer is faced with the decision of how to map these components to the threads. One strategy is to create a thousand threads (one thread for each component) to run on whatever number of processors available. An obvious advantage of this strategy is that it can be easily coded. In this case, however, the overhead time required for creating the threads tends to increase. Its effect becomes more significant when the computational time required for processing the application is small. In addition, with such a relatively large number of threads running on a few processors, a processor is bound to be assigned more than one thread. Consequently, some overhead time is spent by the processor in 'context-switching' (a processor switching from one thread to the other). The net result is a slow-down in the execution of the application. On the other hand, one can map the concurrent components to a number of threads smaller than the number of available processors. This will leave some of the processors idle at some instances of the program execution. As a result, the load balance among the processors is not maintained and the system is not utilized to its fullest capacity.

A third strategy is to code the application program in such a way that the number of concurrent components is mapped to a number of threads equal to the number of available processors. An efficient mapping will balance the number of components mapped to each thread as evenly as possible. Obviously, this strategy can not be as easily coded as the first one. However, compared with the first strategy, fewer threads are created in this case to execute longer jobs. In effect, the overhead time required for creating threads is smaller and less significant. A smaller number of threads will also reduce the effect of context-switching. The net result is a faster concurrent application. In this chapter, two different mapping strategies are investigated and compared. In the first, the number of concurrent components is mapped to an equal number of threads normally greater than the number of available processors. In the second, the number of components is mapped to a number of threads equal to the number of available processors. Comparison between these two strategies is presented in Section 2.10.

2.9 PARALLEL C PROGRAM PASTRANC

The C programming language is used for developing the parallel structural analysis code PASTRANC (PArallel STRuctural ANalysis in C). At this point, the program consists of six 'functions' as shown in

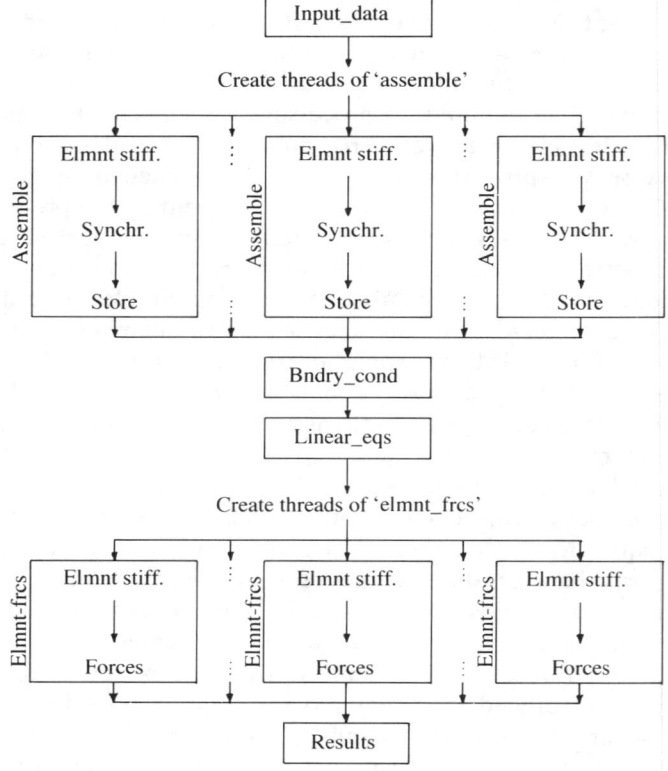

Fig. 3. Functions of the parallel C program (PASTRANC).

Fig. 3. The 'input_data' function reads in and generates input data for different sections, nodal points, and elements. The function 'assemble' sets up the elements stiffness matrices and load vectors due to initial stresses, and then assembles them into the structure (global) stiffness matrix and load vector, respectively. The function 'bndry_cond' adds the external loads to the load vector and applies the boundary conditions to the set of simultaneous equilibrium equations. The function 'linear_eqs' solves the set of reduced simultaneous linear equations for the generalized displacements vector. The function 'elmnt_frcs' sets up the element stiffness matrix for each member in the structure and then calculates the element's generalized force vector and stress vector using eqns (2) and (3), respectively. Finally, the function 'results' prints out the nodal displacements, element forces, and element stresses.

The purpose of the initial experimentation was to explore and demonstrate the concurrent processing capability of Encore Multimax through the use of threads. At this point, we choose to parallelize only certain portions of a serial code, rather than developing parallel algorithms for the entire structural analysis problem. Figure 3 shows portions of the code that are readily concurrent in nature and can be readily parallelized using the Encore Parallel Threads. In function 'assemble', the process of calculating an element stiffness matrix and then assembling it into the global stiffness matrix is independent of the same process performed on other elements. The same argument is valid for function 'elmnt_frcs', where setting up an element stiffness matrix and calculating element forces and stresses of an element can be done completely independently of other elements. Therefore, the two functions 'assemble' and 'elmnt_frcs' are our target for parallelization in this chapter. When function 'assemble' is parallelized, a racing condition arises during assembling the structure stiffness matrix. At this point, a synchronization scheme is included to avoid time-dependent problems and guarantee correct results.

Specifically, our objective in this chapter is to study the following:

1. Overhead time required for creating threads.
2. Comparison of speed-up using monitors versus semaphores.
3. Effect of different mapping strategies on the speed-up expected from the Encore Multimax.
4. Comparison of overall computational time performance of PASTRANC using different options.

5. Effect of amount and frequency of shared memory access (reads from and writes into) on the speed-up expected from the Encore Multimax.

In order to perform this study the following two different sets of options are provided in PASTRANC:

1. Mapping strategies may be selected such that the number of threads (NT) is equal to

    (a) number of elements in the structure (NE), or
    (b) number of available processors (NP).

2. Synchronization mechanisms may be selected as

    (a) semaphores (SEM), or
    (b) monitors (MON), or
    (c) no synchronization, and the processors are allowed to race (RACE); this option guarantees correct answers in the case of one processor, since no racing condition arises; for more than one processor, the results may be incorrect owing to time-dependent problems, as discussed previously in the chapter.

2.10 APPLICATIONS

This section presents a case study of two space trusses: a geodesic dome structure (Fig. 4) and a 72-bar space truss (Fig. 5). The pattern of numbering of nodes and elements of each structure is also shown on these figures. The geodesic dome structure has 61 nodes, 156 elements, 183 degrees of freedom, and bandwidth of 30. The 72-bar space truss has 20 nodes, 72 elements, 60 degrees of freedom, and bandwidth of 24.

As pointed out earlier, only two of the functions of program PASTRANC are parallelized in this preliminary investigation, namely, 'assemble' and 'elmnt_frcs'. Therefore, emphasis is given to these two specific functions. Both functions are parallelized using either NT = NE or NT = NP mapping options. In addition, since function 'assemble' is liable to a racing condition, it is supplied with the following synchronization options: SEM, MON, or RACE.

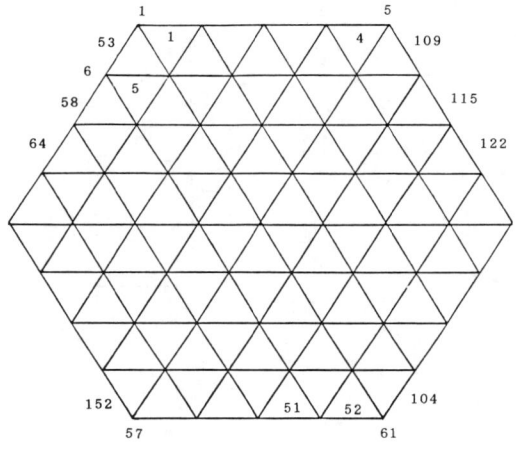
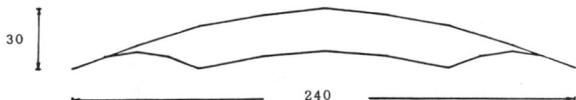

Fig. 4. Geodesic dome structure.

Also, we need to define the notion of 'speed-up' and 'relative speed-up' as used in this work. As an example, the 'speed-up' achieved for function 'assemble' in the case of NP processors is equal to the ratio of the time spent by the sequential code on executing the function 'assemble' to the time spent by the concurrent code on executing the function 'assemble' when NP processors are available. The 'relative speed-up' achieved for function 'assemble' in the case of NP processors is equal to the ratio of the time spent by the concurrent code on executing the function 'assemble' when only one processor is in use to that when NP processors are available.

The Encore Multimax available to the authors at The Ohio State University has 12 processors, but only 11 processors are available for general users. Therefore, all the applications in this chapter (as well as the remaining chapters) are limited to a maximum number of 11 processors. With this in mind, the following subjects are studied: percentage of concurrent calculations, overhead for creating threads,

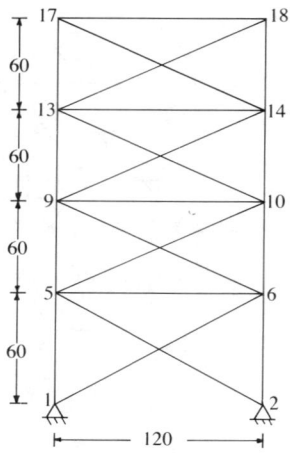

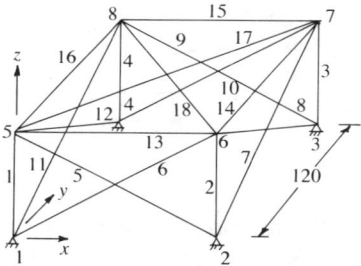

Fig. 5. 72-bar space truss.

speed-up with semaphores versus monitors, speed-up versus elements-thread mapping, relative speed-up, and overall time performance.

2.10.1 Percentage of concurrent calculations

If we exclude the computational time spent on executing I/O functions, namely, 'input_data' and 'results', the remaining time will be spent on executing the following four major computational functions of the structural analysis program: 'assemble', 'bndry_cond', 'linear_eqs', and 'elmnt_frcs'. The pie charts shown in Figs 6 and 7 show the percentages of computational time spent on each of those

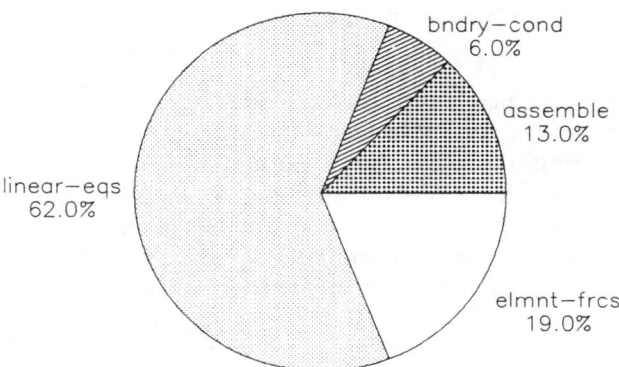

Fig. 6. The computational time pie chart for the geodesic dome structure.

functions by the sequential program, for the geodesic dome structure and the 72-bar space truss, respectively. For the geodesic dome structure, the time spent on computations that are concurrent in nature ('assemble' plus 'elmnt_frcs') amounts to 32% of the total time, while it is equal to 50% for the 72-bar space truss. It is worthwhile to note that the computational time for the function 'linear_eqs' tends to dominate as the size of the structure (number of degrees of freedom and bandwidth) increases.

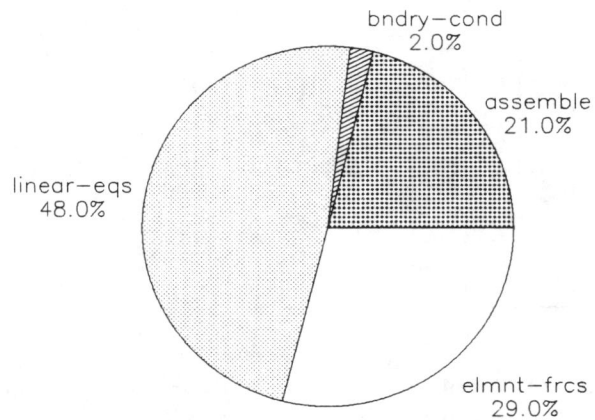

Fig. 7. The computational time pie chart for the 72-bar space truss.

2.10.2 Overhead for creating threads

As mentioned earlier, a thread is created at a low cost. In fact, it takes just under a millisecond (900 μs on the average) to create a thread on the Encore Multimax. Also, it is important to mention that threads are created sequentially. Consequently, the overhead time required for creating the threads is added sequentially to the execution time of the application. In other words, the concurrent processing of the thread starts after the thread has been created, not while it is being created.

Figure 8 demonstrates the effect of the overhead time required for creating threads on the speed-up attained in function 'elmnt_frcs' with option NT = NP for the geodesic dome structure and the 72-bar space truss. The speed-up for the case of one processor is practically unity for both examples. This shows that the overhead time required for creating a thread is negligible compared with the overall execution time when one processor is in use. For more than one processor, however, more threads are created, and the linear speed-up relationship in the number of processors is not maintained. The recorded

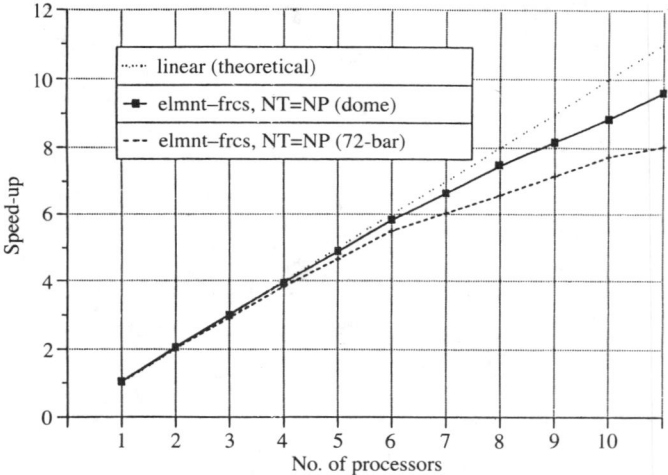

Fig. 8. Effect of threads creation overhead on the speed-up (function 'elmnt_frcs' with NT = NP).

speed-up results show a good match with the formula

$$(speed\_up)_{NP} = \frac{seq\_time}{\frac{seq\_time}{NP} + (NP \times thread\_time)} \quad (4)$$

where $(speed\_up)_{NP}$ is the speed-up attained in function 'elmnt_frcs' with NT = NP option when NP processors are in use, seq_time is the time spent by the sequential code on executing function 'elmnt_frcs', and $thread\_time$ is the overhead time required for creating a thread. For a given application, the only variable in eqn (4) is the number of processors, NP.

The formula (4) and results show that the effect of the overhead time required for creating the threads on the speed-up is less significant when this time is small relative to the sequential time. Figure 8 shows that a better speed-up is achieved for the geodesic dome problem. This is explained by the fact that the sequential time in the geodesic dome is larger than that of the 72-bar space truss, because the former structure has more elements. This causes the effect of the overhead time required for creating threads on the speed-up to be less

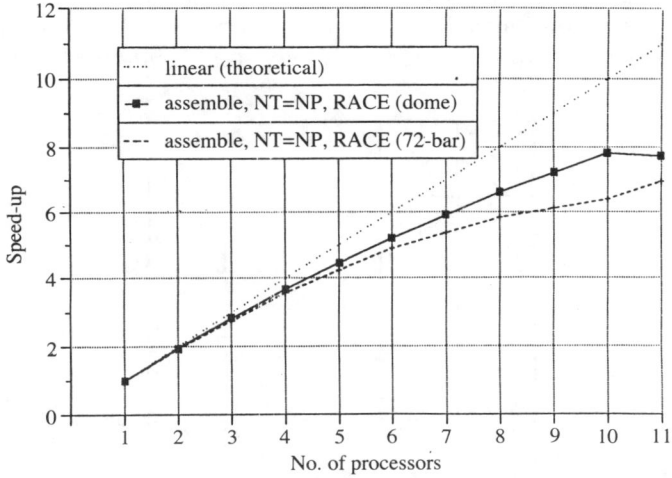

Fig. 9. Effect of threads creation overhead on the speed-up (function 'assemble' with NT = NP and RACE).

significant for the geodesic dome problem. The same discussion and arguments are valid for Fig. 9, which compares the speed-up performance for function 'assemble' with options NT = NP and RACE for the geodesic dome structure and the 72-bar space truss. A sample code that shows the EPT function calls required for the creation of threads is included in the Appendix.

2.10.3 Speed-up with semaphores versus monitors

When implementing any one of the two synchronization mechanisms, semaphores or monitors, the programmer has to be concerned about the efficiency of the implementation. For instance, to avoid a racing condition during assembling the structure stiffness matrix, the programmer can use a single semaphore to synchronize adding up numbers to any location of the stiffness matrix. This means that when one processor is adding up a number in location y of the matrix and another processor is adding up a number in location z, they have to perform their additions mutually exclusive in time. Obviously, this would degrade the speed-up significantly. Moreover, not all the locations of the structure stiffness matrix represent critical sections, and consequently need not be protected against a racing condition.

The programmer needs to safeguard only those locations of the stiffness matrix that may face a racing condition. Figure 10 outlines a typical structure stiffness matrix, where the locations labeled 'X'

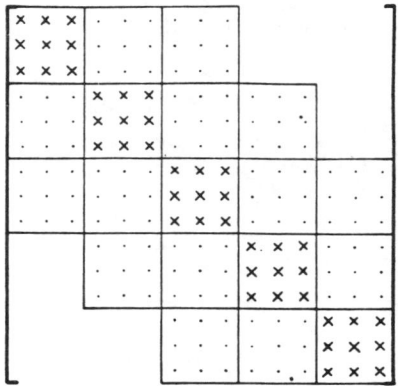

Fig. 10. Structure stiffness matrix (original form).

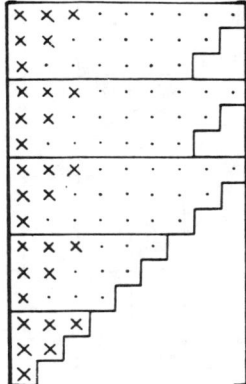

Fig. 11. Structure stiffness matrix (banded form).

represent the locations of critical sections for a typical space truss problem. The total number of these locations is $9 \times num\_of\_nodes$ (nine times the number of nodes in the structure). Given the fact that the structure stiffness matrix is symmetric and banded, it is stored as shown in Fig. 11. This reduces the number of critical sections to $6 \times num\_of\_nodes$. Therefore, the strategy used in this work is to develop a vector of semaphores with dimension equal to the number of critical sections in the structure stiffness matrix. Each element of the semaphore vector guards a corresponding location in the structure stiffness matrix in order to assure that access of different processors to this location during the assembly process occurs mutually exclusive in time. This way, synchronization comes into play only when there is a need to overcome a racing condition. The same concept is adopted when the monitor option is in effect.

Figures 12 and 13 compare the speed-up achieved in function 'assemble' with option NT = NP when using semaphores and monitors for the geodesic dome structure and the 72-bar space truss, respectively. Results show that the execution time for one processor is higher than the sequential time, resulting in a speed-up less than one. This is due to the fact that there is an overhead time associated with the invocation of the synchronization mechanism: semaphores or monitors. The recorded speed-up results show a reasonable match with the

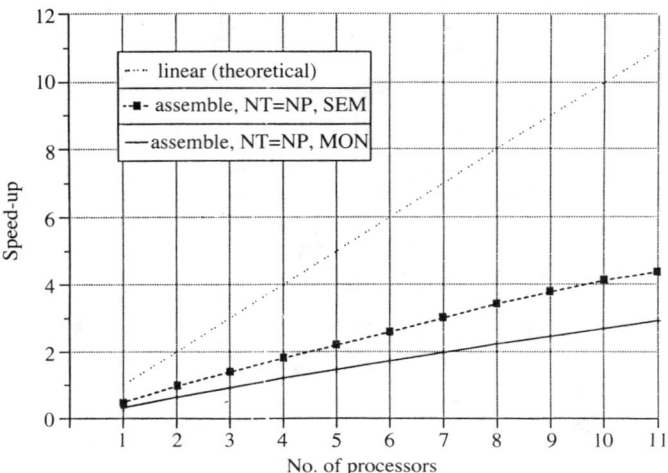

Fig. 12. Speed-up with semaphores versus monitors (geodesic dome structure).

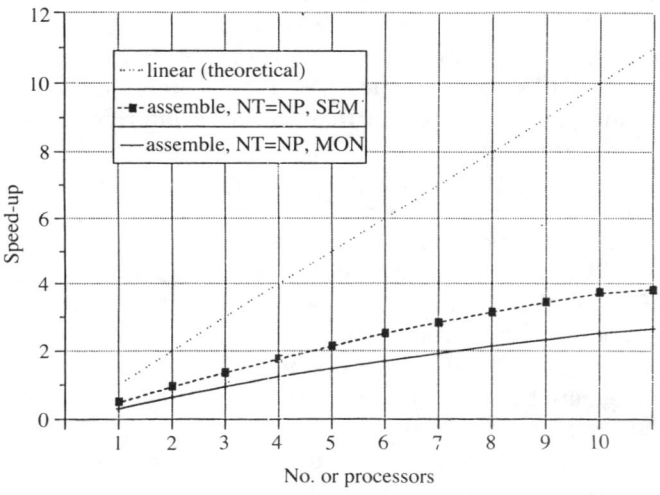

Fig. 13. Speed-up with semaphores versus monitors (72-bar space truss).

formula

$$(speed\_up)_{NP} = \frac{seq\_time}{\frac{seq\_time}{NP} + \frac{syn\_invoc\_time}{NP} + (NP \times thread\_time)}$$

(5)

where $(speed\_up)_{NP}$ is the speed-up attained in function 'assemble' with NT = NP and SEM (or MON) options when NP processors are in use, seq_time is the time spent by the sequential code on executing function 'assemble', syn_invoc_time is the overhead time required for the invocation of semaphores (or monitors), and $thread\_time$ is the overhead time required for creating a thread. For a given application, the only variable in eqn (5) is the number of processors, NP.

The formula (5) and results show that the overhead time required for the invocation of the semaphores is less than that required for the monitors. This is due to the fact that the latter mechanism is more structured than the former. Results also show that the effect of the overhead time required for the invocation of the synchronization mechanism tends to decrease as the number of the processors increases. A sample code that shows the EPT function calls required for synchronization using semaphores is included in the Appendix.

2.10.4 Speed-up versus elements-thread mapping

Two different strategies for mapping are considered in this chapter. First, a number of threads (NT) equal to the number of elements in the structure (NE) is developed for each of the functions 'assemble' and 'elmnt_frcs' in program PASTRANC. In other words, each thread will have only one element mapped to it. Therefore, a fixed number of threads is created, irrespective of the number of the processors available. In the second strategy, a number of threads (NT) equal to the number of available processors (NP) is developed for each of the functions 'assemble' and 'elmnt_frcs'. In this case, the elements of the structure are mapped to the processors as evenly as possible.

Figures 14 and 15 show the speed-up performance in cases of NT = NE mapping versus NT = NP mapping for the function 'elmnt_frcs' for the geodesic dome structure and the 72-bar space truss, respectively. Similar results for the function 'assemble' are shown in Figs 16 and 17. In all of these cases, the speed-up for the case

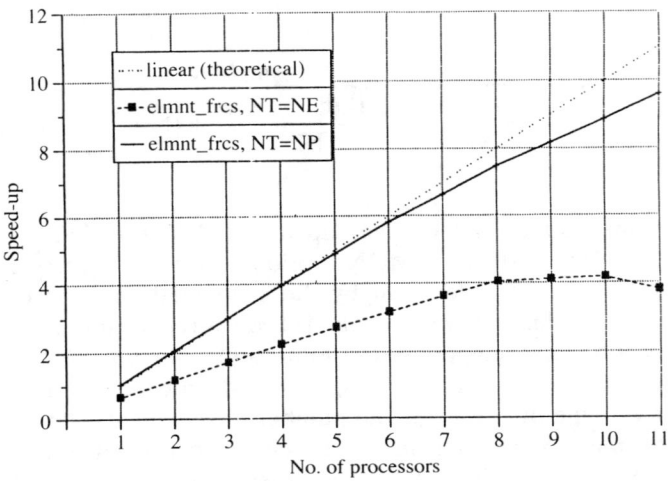

Fig. 14. Speed-up with elements-thread mapping for the function 'elmnt_frcs' (geodesic dome structure).

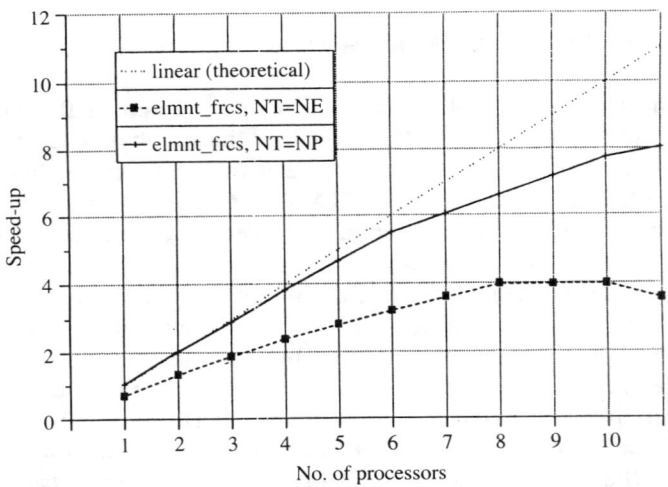

Fig. 15. Speed-up with elements-thread mapping for the function 'elmnt_frcs' (72-bar space truss).

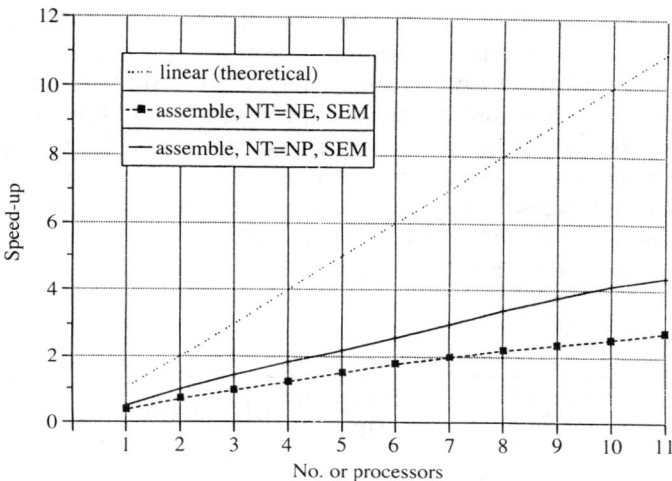

Fig. 16. Speed-up with elements-thread mapping for the function 'assemble' (geodesic dome structure).

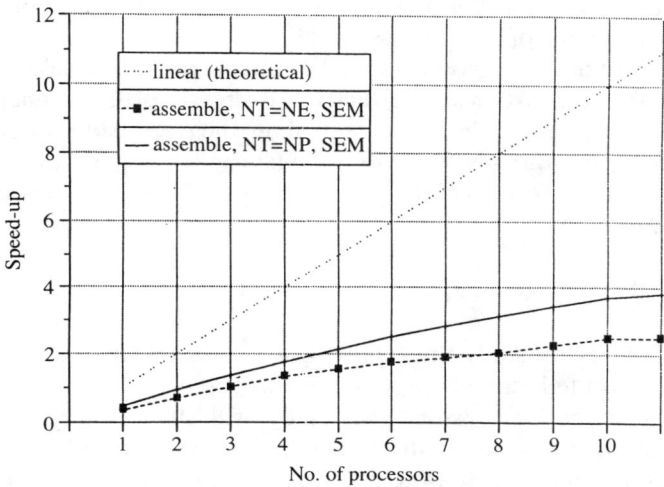

Fig. 17. Speed-up with elements-thread mapping for the function 'assemble' (72-bar space truss).

NT = NE is worse than that for the case NT = NP. This is due to the large number of threads created in the former case. As discussed earlier, the overhead time required for creating threads slows down the solution process and affects the speed-up. In addition, with more than one thread assigned to each processor, more time is spent in context-switching (a processor switching from one thread to the other) within each processor. This makes the application even slower. In all the cases with the NT = NE option, a reasonable match is found between the results and the formula

$$(speed\_up)_{NP} = \frac{seq\_time}{\frac{seq\_time}{NP} + \frac{switch\_time}{NP} + \frac{syn\_invoc\_time}{NP}(if any) + (NE \times thread\_time)} \quad (6)$$

where $(speed\_up)_{NP}$ is the speed-up attained in the function with the NT = NE option when NP processors are in use, seq_time is the corresponding time spent by the sequential code on executing the function, $switch\_time$ is the overhead time due to context-switching, syn_invoc_time is the overhead time required for the invocation of semaphores (if any), and $thread\_time$ is the overhead time required for creating a thread. For a given application, the only variable in eqn (6) is the number of the processors, NP.

The formula (6) shows that the effect of the overhead time consumed in context-switching tends to decrease as the number of the processors increases. The results also show the overhead time required for creating the threads tends to dominate the execution time for the function 'elmnt_frcs' with the NT = NE option, in the cases of 9, 10, and 11 processors.

2.10.5 Relative speed-up

It is of interest to study the relative gain achieved when a concurrent code is executed on NP processors compared with the same code executed on one processor. This study will show the effect of the overhead times due to creation of threads, invocation of semaphores or monitors, and context-switching (which are taxing in the case of one processor) when the application is executed on more than one processor.

Figure 18 and 19 show the relative speed-up using two different

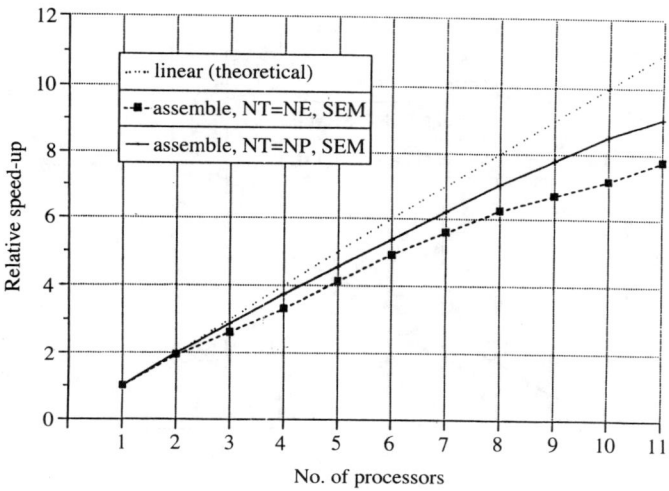

Fig. 18. Relative speed-up with elements-thread mapping for the function 'assemble' (geodesic dome structure).

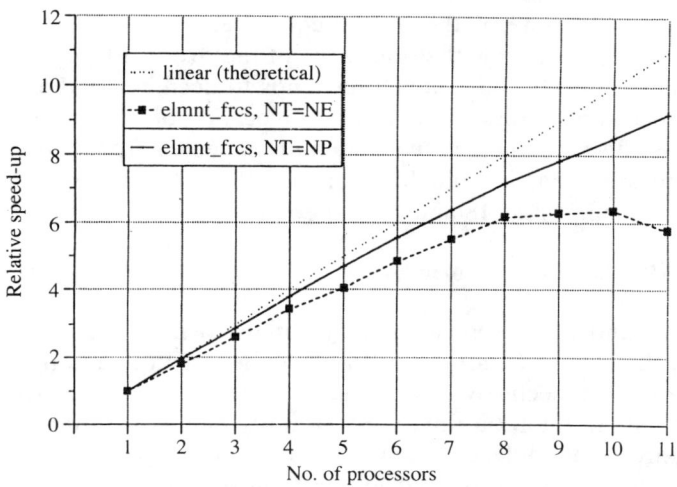

Fig. 19. Relative speed-up with elements-thread mapping for the function 'elmn_frcs' (geodesic dome structure).

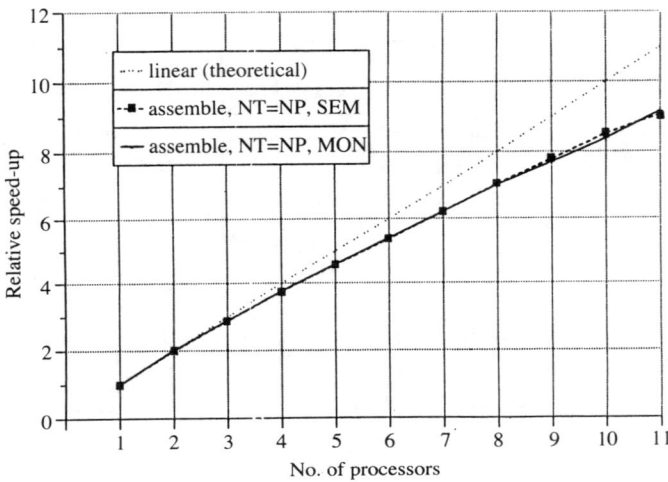

Fig. 20. Relative speed-up with semaphores and monitors (geodesic dome structure).

mapping strategies for the functions 'assemble' and 'elmnt_frcs', respectively. These results (as well as eqn (4)), show that the larger the number of threads, the worse is the relative speed-up, since the overhead time required for creating the threads is added sequentially to the execution time of the concurrent code. Figures 18 and 20, in addition to eqn (5), show that the effect of the overhead time required for the invocation of the semaphores tends to decrease as the number of the processors increases. Therefore, invocation of semaphores has a negligible effect on the relative speed-up expected from the system. The same argument is valid for monitors (Fig. 20 and eqn (5)) and context-switching (Figs 18 and 19 and eqn (6)).

2.10.6 Overall time performance

Figures 21 and 22 present a summary of the overall performance of program PASTRANC, for the geodesic dome structure and the 72-bar space truss, respectively. The ratio of the total time spent in PASTRANC to the total time spent in the sequential (serial) solution is compared for different numbers of processors using different options. The horizontal dashed line passing through 1 on the vertical scale represents the datum for the time spent in the sequential

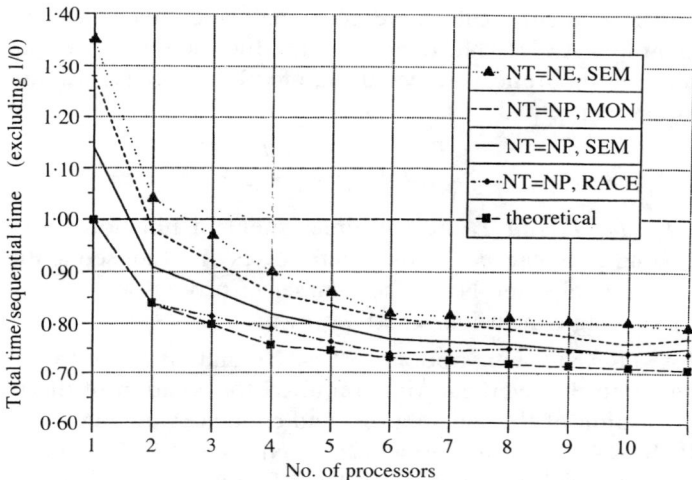

Fig. 21. Overall time performance of PASTRANC for the geodesic dome structure.

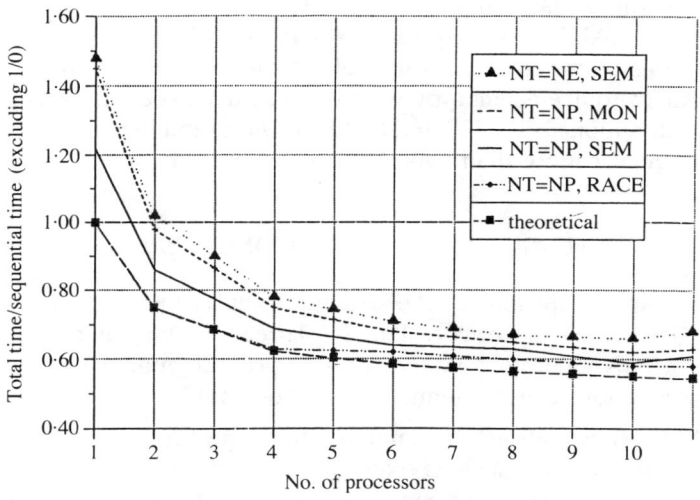

Fig. 22. Overall time performance of PASTRANC for the 72-bar space truss.

solution. The 'theoretical' curve represents the theoretical expected time spent by PASTRANC divided by the time spent by the sequential solution. The theoretical expected time for PASTRANC running with NP processors is equal to

$$\frac{T_1}{NP} + T_2 + T_3 + \frac{T_4}{NP}$$

where T_1, T_2, T_3 and T_4 are the times spent in functions 'assemble', 'bndry_cond', 'linear_eqs', and 'elmnt_frcs' in the sequential program, respectively, and NP is the number of processors.

The following conclusions can be drawn:

1. PASTRANC with options NT = NE and SEM is the slowest, owing to the overhead time required for creation of the threads, invocation of the semaphores, and context-switching.
2. PASTRANC with options NT = NP and SEM is faster than PASTRANC with options NT = NE and SEM, since the overhead time required for creating threads and the effect of context-switching is smaller in the former case.
3. PASTRANC with options NT = NP and MON is slower than PASTRANC with options NT = NP and SEM, since the time required for invocation of monitors is higher than that required for semaphores. This is due to the fact that monitors are more structured than semaphores.
4. PASTRANC with options NT = NP and RACE is the fastest, yet it guarantees correct results only in the case of one processor.
5. An overall efficiency of 90–95% was achieved in PASTRANC with options NT = NP and SEM for the example problems solved whenever more than two processors are used.

2.11 SUMMARY AND CONCLUSIONS

In this chapter, portions of the structural analysis problem have been parallelized using the Encore Parallel Threads (EPT) that operate on top of UNIX operating system and C programming language. The main conclusions can be summarized as follows:

1. In this application, the amount and frequency of shared memory access requests (such as reads or writes) do not have a significant effect on the speed-up expected from the system. In other words, the contention of the bus has an insignificant effect.

2. In this application, the overhead time due to synchronization (not that due to invocation of semaphores or monitors) seems to be negligible. This is due to the fact that the critical sections are very short in length (computationally small) and not too many in number.
3. In a concurrent code, there is an overhead time due to creation of threads, invocation of the synchronization mechanism, and context-switching. The effect of these factors on the speed-up tends to be less significant when the overhead time is small compared with the execution time of the corresponding sequential code.
4. The speed-up and relative speed-up deteriorate as the number of threads increases. This is particularly true for small problems.
5. The overhead time required for the invocation of semaphores affects the speed-up performance of the system. However, it has a negligible effect on the relative speed-up. The same argument is valid for monitors, with the addition that they have a more significant effect on the speed-up since the time required for their invocation is larger. In general, semaphores are used where simple mutual exclusions exist and speed is of primary concern, while monitors are more reliable to use when exceptions are involved.
6. The overhead time required for context-switching affects the speed-up performance of the system. However, it has a negligible effect on the relative speed-up.
7. An overall efficiency of 90–95% is achieved for more than two processors when the number of threads is equal to the number of the processors and semaphores are used for synchronization.

This chapter has demonstrated the feasibility of parallel processing in the area of structural analysis through the use of the notion of cheap concurrency and the concept of threads.

Chapter III

Automatic Partitioning of Framed Structures for Concurrent Processing

3.1 INTRODUCTION

A strategy that is usually used in connection with multiprocessor computers is the 'divide and conquer' paradigm. The underlying concept is to partition a given task into several smaller ones, which are then assigned to various processors (Fig. 23). In this chapter, we present a three-stage algorithm for partitioning the domain of a given structure into several smaller subdomains in a process known as 'partitioning', 'substructuring', or 'subdomaining'. As an example, Fig. 24 shows the results of applying a partitioning algorithm to a four-story three-bay plane frame structure, for the cases of one, two, three, and four processors.

Noor et al. [1978] reviewed the static partitioning techniques and their appliction to structural analysis problems. They focused on several aspects including multilevel partitioning, use of hypermatrix, and other sparse matrix schemes. Farhat et al. [1987b] presented a partitioning technique for finite element analysis that results in strong decoupling between subdomain matrices, thus leading to considerable computational savings in terms of concurrent computations. Noor & Peters [1989] combined partitioning techniques with operator splitting and iterative procedures to solve nonlinear finite element analysis problems on multiprocessor computers.

Generally, the major steps involved in the finite element analysis of structures can be summarized as shown in Fig. 1. If we exclude the I/O (input/output) functions, the five major computational steps are

1. Assembling the structure stiffness matrix;

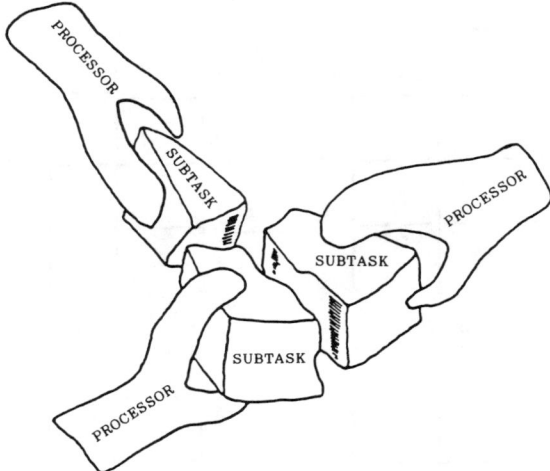

Fig. 23. The divide and conquer paradigm.

2. Assembling the load vector;
3. Applying the boundary conditions;
4. Solving for the nodal generalized displacements;
5. Calculating the element forces and stresses.

A possible intuitive approach is to attempt to parallelize each of the previous functions in some appropriate manner. For example, steps 1 and 5, namely, assembling the stiffness matrix and calculating the element forces and stresses, can be parallelized through mapping the elements of the structure to a number of sets equal to the number of the available processors. An efficient mapping should balance, as evenly as possible, the number of elements assigned to each set and consequently the load among the processors. In Chapter II, we implemented this strategy on the Encore Multimax through modification of a sequential code, and have found speed-ups of up to 90% (Adeli & Kamal [1989a, b]).

Similarly, this logic can be extended to include steps 2 and 3, namely, assembling the load vector and applying the boundary conditions. For example, step 2 can be parallelized through mapping an equal (or as equal as possible) number of nodes to each processor. Similarly, the restrained (support) degrees of freedom can be mapped as evenly as possible to the processors. In this approach, each step is

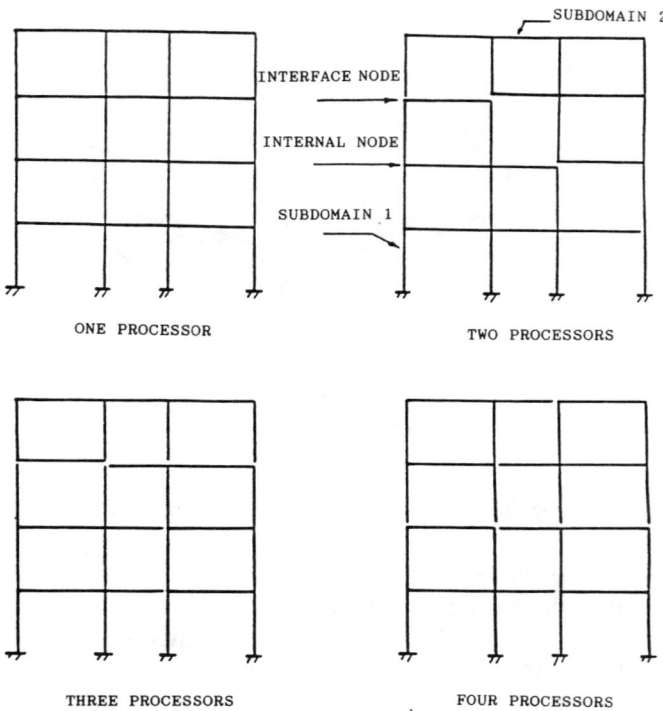

Fig. 24. Domain partitioning for a four-story three-bay frame.

parallelized in a manner different from other steps. It is worthwhile noting that this approach is more efficiently applied to multiprocessor computers with shared memory such as the Encore Multimax (Encore [1985]) than those with no shared memory such as the Intel hypercube. This is due to the fact that the shared memory in the former is accessible by all the processors at all times; consequently a high degree of flexibility is provided, and the programmer can choose the more appropriate discretization or mapping pattern or scheme for each step of the solution process.

This approach, however, cannot be easily extended to include the remaining step 4, since the solution of the simultaneous linear equations is inherently sequential. Many algorithms have been developed for parallelizing this step (see e.g. Adams [1985] and Saad [1986]). However, this step still represents a bottleneck for the efficient concurrent processing of structural analysis problems. Farhat *et al.* [1987a] showed that the effect of 'unparallelizable' code on degrading the speed-up expected from the system is substantial, even when the percentage of this portion in the whole code is small.

Let the total time spent on executing the sequentional code be *parallel_time* + *nonparallel_time*, where *parallel_time* represents the execution time spent on the parallelizable portion of the code, while *nonparallel_time* represents the time spent on executing the non-parallelizable portion. Then, speed-up for *NP* processors ($speed\_up_{NP}$), according to Amdahl's second law, will be

$$speed\_up_{NP} = \frac{parallel\_time + nonparallel\_time}{\dfrac{parallel\_time}{NP} + nonparallel\_time}$$

This equation shows that the larger the *nonparallel_time*, the greater will be the deviation from the linear speed-up case. This offset increases with the number of processors, *NP*, used. As an example, if the *parallel_time* represents 95% of the total time then the *nonparallel_time* will represent 5% of the total time, and a maximum speed-up of 20 is attained when an infinite number of processors is in use. Thus the maximum processing capability of the system clearly has not been utilized. In addition, the time required for solving the set of linear equations tends to substantially dominate the total execution time for sequential codes as the size of the structure (number of degrees of freedom and bandwidth) increases. This brings the idea of directly parallelizing an existing sequential code for structural analysis to a halt. Consequently, a different strategy is adopted in this book.

The rest of this chapter is organized as follows. Section 3.2 outlines some basic concepts and definitions. Then, the vectors and matrices required for applying the algorithm are identified in Section 3.3. The generation of the initial subdomains, and their adjustment into the intermediate subdomains are described in Sections 3.4 and 3.5. The final adjustment, resulting in the final subdomains, is given in Section 3.6, and the 'post-partitioning' in Section 3.7. Throughout these sections, the concepts discussed are demonstrated on the example frame structure shown in Fig. 25. In this example, we limit attention to the case of three processors for the sake of illustration only. The pattern of numbering of the nodes and the elements of this frame is also shown in Fig. 25. Section 3.8 illustrates the results of the C

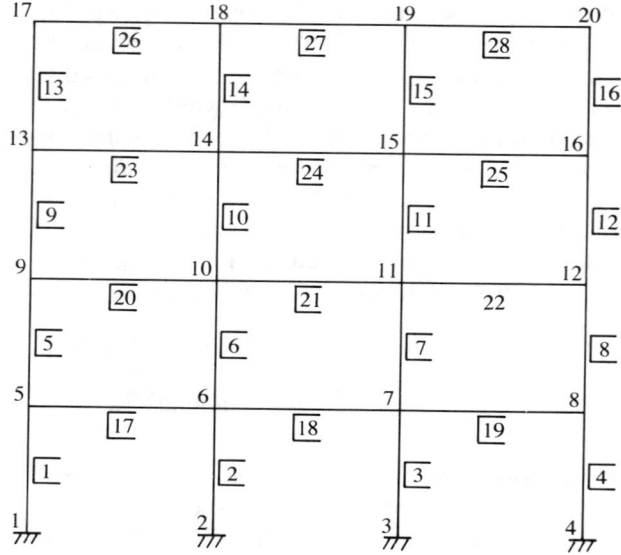

Fig. 25. Pattern of numbering for the four-story three-bay frame.

program coded for the algorithm when applied to several plane and space frame structures, for a variable number of processors. Finally, in Section 3.9, some conclusions are drawn for the partitioning algorithm developed in this chapter.

3.2 BASIC CONCEPTS AND DEFINITIONS

Rather than attempting to parallelize the solution process on the local level (the steps level), we attempt to accomplish this on the global level (the structure level). Basically, we rearrange the input data through a partitioning approach so that subsequent analysis and design steps can take place as efficiently as possible, utilizing the general architecture of the multiprocessor computers. We choose to partition the domain into a number of subdomains equal to the number of prescribed processors. This mapping strategy (matching the numbers of processors and subdomains) proved to be more efficient than other strategies (Adeli & Kamal [1989a, b]). Once the domain has been partitioned in this manner, the nodes in the structure are classified as 'internal nodes', 'interface nodes', and 'support nodes'. An internal node or a closed node is a node whose connected elements belong to one subdomain only. An interface node, a border node, or an open node is a node whose connected elements belong to more than one subdomain. A support node or a boundary node is a node with all of its displacement degrees of freedom restrained. Supports are not classified as internal nodes or interface ones.

We use the following numbering scheme. The internal nodes of the subdomains and the support nodes entirely contained within the subdomains are numbered first, followed by the interface nodes between the subdomains and any support node falling on the borders of the subdomains. This numbering scheme results in a structure stiffness matrix of the form shown in Fig. 26. The submatrix $K_{II}(q)$ and the vectors $r_I(q)$ and $R_I(q)$ represent the stiffness matrix, generalized displacements, and load vectors associated with subdomain q (SD_q), respectively. The submatrix $K_{IB}(q)$ represents the stiffness matrix coupling the subdomain q (SD_q) and the interface (SD_B).

As an example, the new numbering scheme and the corresponding stiffness matrix for the four-story three-bay frame and the case of three processors is shown in Fig. 27. The non-cross-hatched small square blocks represent the nonzero terms of the stiffness matrix. The cross-hatched small square blocks represent the degrees of freedom associated with the support nodes. These degrees of freedom are usually ignored during the solution of the linear equations; therefore, the rows and columns associated with these degrees of freedom can be dropped from the stiffness matrix at this point. The large blank boxes represent the zero terms. The structure of the stiffness matrix shown in Fig. 26 is suitable for distribution among the processors. Moreover, the strong decoupling of the diagonal blocks $K_{II}(q)$ shown in Fig. 26 suggests that they can be processed concurrently and their effect can be collapsed onto the interface matrix K_{BB}. Thus, in this approach, the only set of simultaneous linear equations that need be rearranged for parallel solution is the one associated with the interface nodes. Clearly, this set is much smaller than the original one, and thus a more tractable situation results.

In the above discussion, a substructuring concept has been described, but not the algorithm itself. In fact, many algorithms can be developed for automatic partitioning of structural domains. However, a successful algorithm must possess certain features. The desirable

Parallel Processing in Structural Engineering

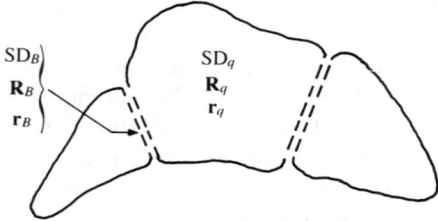

SD Subdomain

R   Load vector

r   Displacement vector

q Subscript denoting subdomain q

B Subscript denoting border (interface)

I Subscript denoting interior

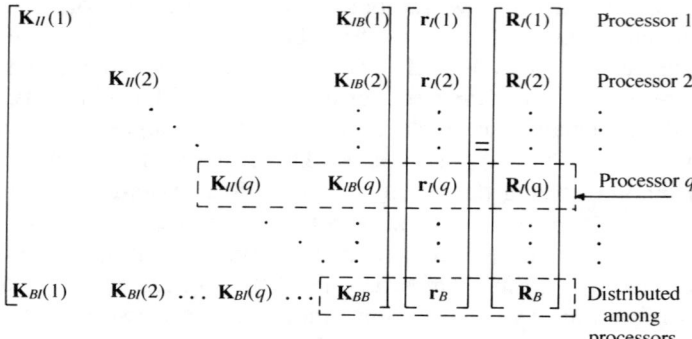

Fig. 26. Mapping of subdomains onto processors.

features should include (Farhat *et al.* [1987a]) the following:

1. Hardware or machine independence—a feature that facilitates portability and flexibility.
2. General purpose—a feature that enables handling irregular domains and irregular patterns of numbering. For example, the partitioning scheme shown in Fig. 28 for the case of three processors results from an algorithm that is sensitive to the

Automatic Partitioning of Framed Structures 39

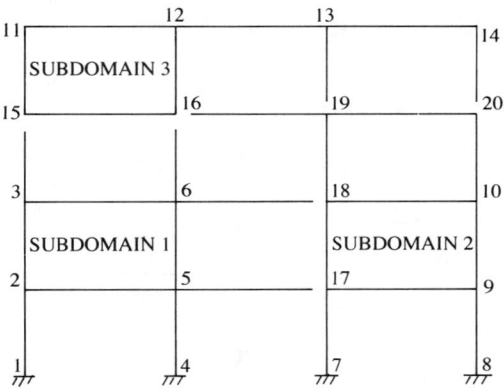

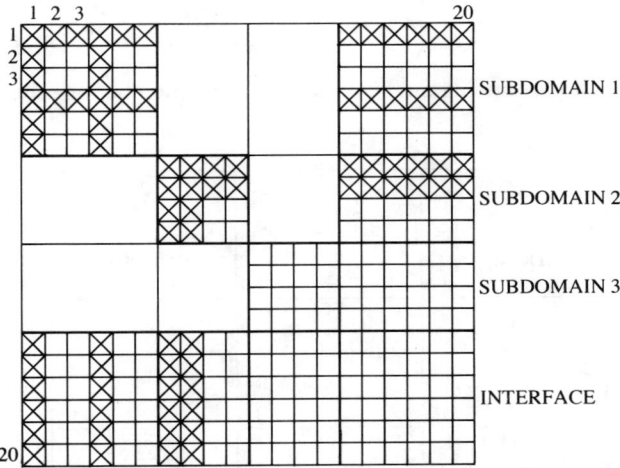

Fig. 27. A partitioning scheme and the corresponding stiffness matrix.

pattern of numbering (i.e. when the numbering scheme changes, a different set of subdomains results). This sensitivity is undesirable since it ties the efficiency of solving the problem to the programmer's style of choosing a numbering scheme. For an algorithm to be insensitive to domain irregularity or pattern of numbering, it has to propagate through the structure based on connectivity rather than the numbering schemes.

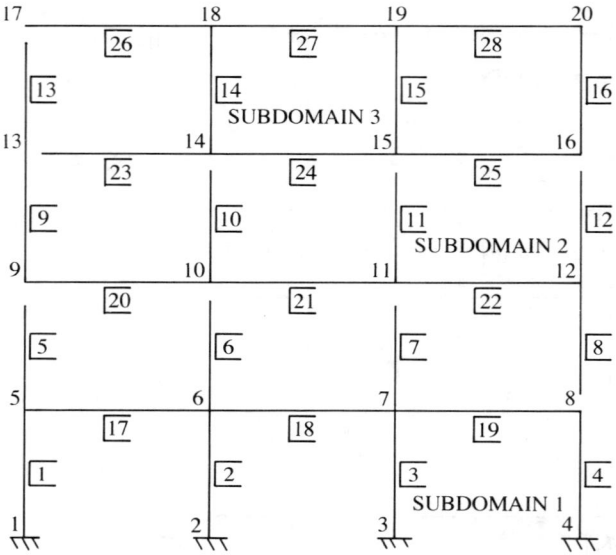

Fig. 28. A partitioning method sensitive to the numbering schemes.

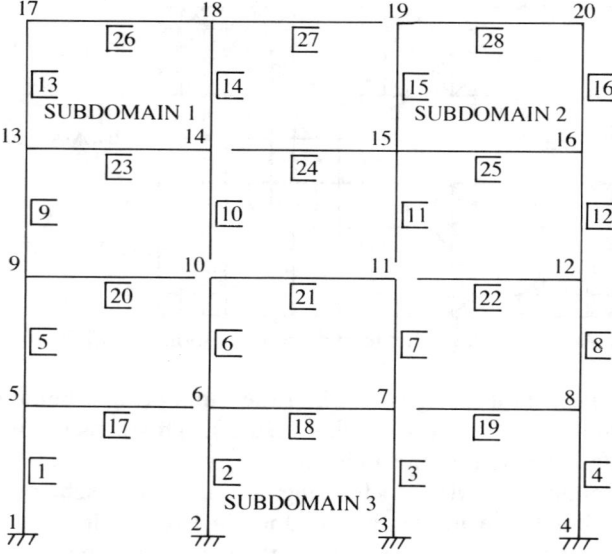

Fig. 29. A case of unbalanced subdomains.

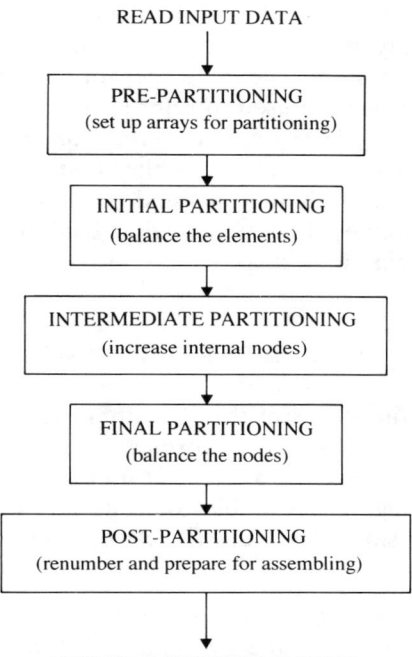

Fig. 30. Macro flow chart of the partitioning algorithm.

3. Workload balance—a feature that maintains the best possible balance of workload among the processors throughout the calculations. For example, Fig. 29 shows a partitioning pattern in which the balance of elements among the processors is not maintained (the number of elements in one of the subdomains is almost one-half the number of elements in the other two). More elaborate discussion on the issue of load balance will be presented in subsequent sections.
4. A small number of interface nodes—a feature that considerably minimizes the number of the simultaneous equations to be solved in parallel, and thus the severity of the bottleneck situation that arises during their parallel solution. For example, the partitioning pattern shown in Fig. 27 results in 6 interface nodes (i.e. a set of 18 linear equations), while the one shown in Fig. 28 results in 9 interface nodes (i.e. 27 linear equations).

To this end, we note that the terms 'parallel' and 'concurrent' are used interchangeably to refer to the simultaneous execution taking place on several processors. The same goes for the terms 'substructure' and 'subdomain', which are used to refer to a portion of the structure that is mapped to a processor. The term *num_nodes* refers to the number of nodes of the structure, while *num_elmnts* refers to the number of elements. With this in mind, our algorithm for domain partitioning is presented in the following sections. Figure 30 summarizes various partitioning stages.

3.3 PRE-PARTITIONING

Before describing the stages of the algorithm, data for the nodes and elements are arranged into certain format. Using the input data given by the user, matrices and vectors are set up for use in the subsequent stages of partitioning. The variables or constants within the brackets indicate the dimension of vectors ([. . .]) or dimensions of matrices ([. . .][. . .]).

1. *elmnts_info*[*num_elmnts*][2]: a matrix that stores the connectivity information (i.e. the two end nodes) associated with each element.
2. *nodal_wghts*[*num_nodes*]: a vector that stores the 'weight' of each node defined as the number of elements attached to it.
3. *elmnts_of_node*[2 × *num_elmnts*]: a vector that stores the elements attached to each node.
4. *wghts_pointer*[*num_nodes*]: a vector whose *i*th location stores the location in the *elmnts_of_node* array where the list of elements attached to the (*i* + 1)th node begins.
5. *other_node*[2 × *num_elmnts*]: a vector that stores in each location the other node connected to the element listed in the same location of the *elmnts_of_node* vector.
6. *istart*[*NP* + 1]: a vector that determines the initial number of elements within each subdomain. This number is particularly important for generating the initial subdomains. The initial number of elements assigned to subdomain $q(SD_q)$ is equal to the difference between the two numbers stored in locations q and $q - 1$ of the *istart* vector.

Table 1 lists all of the above-mentioned matrices and vectors for the

TABLE 1
Characteristic arrays for the example frame of Fig. 25 (case of 3 processors)

No.	nodal_wghts	wghts_pointer	elmnts_of_node	other_node	istart
0	1	0	1	5	0
1	1	1	2	6	10
2	1	2	3	7	19
3	1	3	4	8	28
4	3	4	1	1	
5	4	7	5	9	
6	4	11	17	6	
7	3	15	2	2	
8	3	18	6	10	
9	4	21	17	5	
10	4	25	18	7	
11	3	29	3	3	
12	3	32	7	11	
13	4	35	18	6	
14	4	39	19	8	
15	3	43	4	4	
16	2	46	8	12	
17	3	48	9	7	
18	3	51	5	5	
19	2	54	9	13	
20			20	10	
21			6	6	
22			10	14	
23			20	9	
24			21	11	
25			7	7	
26			11	15	
27			21	10	
28			22	12	
29			8	8	
30			12	16	
31			22	11	
32			9	9	
33			13	17	
34			23	14	
35			10	10	
36			14	18	
37			23	13	
38			24	15	
39			11	11	
40			15	19	

(continued)

TABLE 1—contd.

No.	nodal_wghts	wghts_pointer	elmnts_of_node	other_node	istart
41			24	14	
42			25	16	
43			12	12	
44			16	20	
45			25	15	
46			13	13	
47			26	18	
48			14	10	
49			26	17	
50			27	19	
51			15	15	
52			27	18	
53			28	20	
54			16	16	
55			28	19	

example frame shown in Fig. 25. The information stored in these arrays does not change during the partitioning stages. Note that array referencing in the C programming language starts from zero rather than 1. This is shown in Table 1 in the left-most column, which determines the location referencing in the arrays.

3.4 INITIAL PARTITIONING

This section presents the algorithm for the initial partitioning of structures consisting of two-node elements. The basic idea of this stage is to distribute, as evenly as possible, the elements of the structure into a number of sets equal to the number of prescribed processors. The following arrays are set up in this stage of partitioning:

1. nodes_active[num_nodes]: a vector that stores the nodes in the order they are stacked during partitioning.
2. elmnts_active[num_elmnts]: a vector that stores the elements in the order they are stacked during partitioning.
3. elmnts_trace[num_elmnts]: a vector that identifies the subdomain in which each element is located.
4. nodes_trace[num_nodes]: a vector that keeps track of the nodes stacked during the partitioning process. After the initial partitioning stage has been completed, this vector is used to identify

whether a node is an internal node, an interface node, or a support node.
5. *subdomains_trace*[2 × *num_elmnts*]: a vector that classifies the elements connected to each node according to the subdomain in which they are located.
6. *statistics*[*NP*][2]: a matrix storing the number of elements and internal nodes corresponding to each subdomain in its first and second columns, respectively. Note that supports are not considered internal nodes or interface nodes, since their corresponding equations are ignored during the process of solving the simultaneous linear equations.

Also, note that the information stored in the *nodes_trace, elmnts_trace, subdomains_trace*, and *statistics* arrays is updated from one stage of partitioning to the next, in contrast to the arrays described in Section 3.3: *nodes_active* and *elmnts_active*.

The step-by-step algorithm for the initial partitioning stage is presented in Table 2 and Fig. 31. This algorithm has been applied to

TABLE 2
Algorithm for initial partitioning

1. Locate the node with minimum weight. This is the first active node. Store it in the first (zero) location of the *nodes_active* array. Set the node counter to 1. Set the subdomain counter to 1.
2. Stack the elements attached to the active node (only those that have not already been stacked) in the *elmnts_active* array. Adjust the subdomain counter (if necessary) using the *istart* array. Label the elements that have been stacked by storing the number of subdomain they belong to in the *elmnts_trace* and *subdomains_trace* arrays. Label the location of the active node in the *nodes_trace* array.
3. If the node counter equals *num_nodes*, the initial partitioning stage is complete; go to step 4. Otherwise, locate the first element in the *elmnts_active* array with an end node not yet stacked in the *nodes_active* array. Stack this node in the *nodes_active* array. The current element is the active element. Increase the node counter by 1 to determine the next active node in the *nodes_active* array; go to step 2.
4. Use the *istart*[1] − *istart*[0] elements in the *elmnts_active* array as the elements of the first subdomain. Similarly, use *istart*[2] − *istart*[1] elements for the second subdomain, and so on.
5. Adjust the *nodes_trace* array such that each location stores the number of the subdomain in which the node is located. Negative values indicate supports that are internal. Zero values indicate interface nodes or supports at interface.

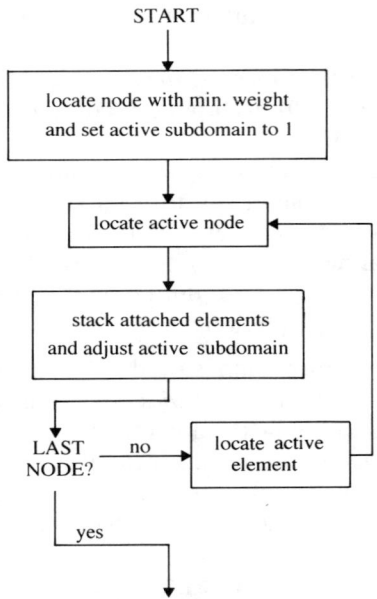

Fig. 31. Flow chart for initial partitioning.

the plane frame structure shown in Fig. 25, for the case of three processors. Table 3 lists the arrays involved in the substructuring process. The resulting partitioned structure and subdomains statistics are shown in Fig. 32. Note that the algorithm starts at a node with minimum weight, which usually resides at one corner or end of the structure. In addition, it goes through the structure based on connectivity, not on the patterns of numbering. This results in a partitioning scheme that is independent of the pattern of numbering of the structure. This makes the algorithm robust and quite general in terms of handling irregular geometries. Furthermore, the algorithm results in subdomains that are compact (i.e. the elements of each subdomain are contiguous). This feature reduces the number of interface nodes and the number of linear equations associated with them.

On the other hand, the algorithm presented so far has the shortcoming of not maintaining a balance of internal nodes among the processors. When the internal nodes and supports located within the subdomains of the partitioned frame shown in Fig. 32 are numbered

TABLE 3
Characteristic arrays at the end of the initial partitioning stage for the example frame of Fig. 25 (case of 3 processors)

No.	nodes_active	elmnts_active	elmnts_trace	nodes_trace	subdomains_trace
0	1	1	1	−1	1
1	5	5	1	−1	1
2	9	17	2	−2	2
3	6	9	3	−3	3
4	13	20	1	1	1
5	10	2	1	1	1
6	2	6	2	0	1
7	7	18	3	0	1
8	17	13	1	1	1
9	14	23	2	0	1
10	11	10	2	0	1
11	3	21	3	3	2
12	8	3	1	1	2
13	18	7	2	0	1
14	15	19	3	0	2
15	4	26	3	3	3
16	12	14	1	0	3
17	19	24	1	0	2
18	16	11	2	3	1
19	20	22	1	3	1
20		4	2		1
21		8	3		1
22		27	1		2
23		15	2		1
24		25	3		2
25		12	2		2
26		28	3		2
27		16	3		2
28					3
29					3
30					3
31					3
32					1
33					1
34					1
35					2
36					2
37					1
38					2
39					2

(continued)

TABLE 3—contd.

No.	nodes_active	elmnts_active	elmnts_trace	nodes_trace	subdomains_trace
40					3
41					2
42					3
43					3
44					3
45					3
46					1
47					2
48					2
49					2
50					3
51					3
52					3
53					3
54					3
55					3

first followed by the interface nodes and supports located on the interface, in order to achieve the favorable decoupling of subdomain matrices shown in Fig. 26, the new numbering scheme and the corresponding stiffness matrix will be as shown in Fig. 33. The balance of load among the processors is not maintained, since there are no subdomain matrices associated with subdomain 2.

The approach of domain partitioning based solely on balancing the number of elements within each subdomain is not entirely efficient. This approach will balance the workload only in certain steps of structural analysis and optimization to be discussed in the following chapters. Therefore, we need to find a partitioning scheme in which the degrees of freedom (or internal nodes) are balanced within the subdomains. This is the main thrust of the rest of this chapter. However, the step of initial partitioning is important, since it provides the program with a good starting point and a reasonable initial estimate of the number of inteface nodes and, hence, the number of internal nodes to be balanced among the subdomains. Therefore, the next two sections are dedicated to adjusting the subdomains resulting from the initial partitioning stage, in an attempt to assign each of the subdomains a 'fair' share of the internal nodes, so that the program can continue efficiently in the subsequent analysis and design steps of the solution process.

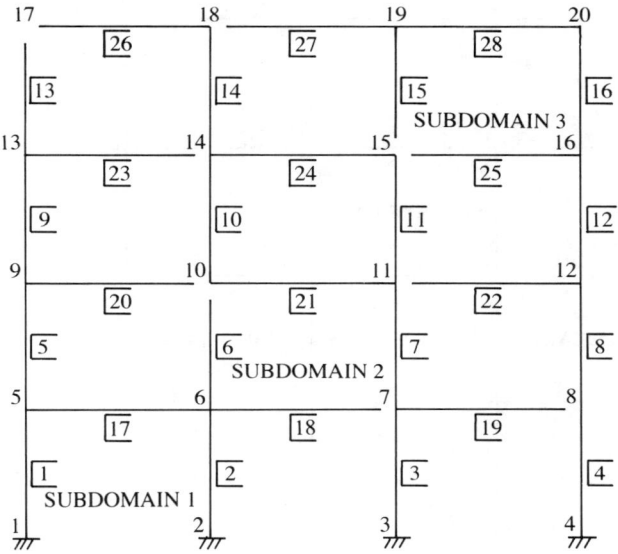

SUBDOMAIN	NUMBER OF INTERNAL NODES	NUMBER OF ELEMENTS
1	4	10
2	0	9
3	4	9

NUMBER OF INTERFACE NODES = 8

Fig. 32. Partitioned structure and domains statistics at the end of the initial partitioning stage for the example frame of Fig. 25 (case of 3 processors).

3.5 INTERMEDIATE PARTITIONING

The purpose of this stage is to attempt to increase the overall number of internal nodes (or decrease the number of interface nodes) attained in the initial partitioning stage. The underlying concept is to scan all the interface nodes and attempt to 'close' any of them. An interface node is switched to an internal node through a process of exchange of elements between subdomains only if during this process no internal node is switched to an interface node. The step-by-step algorithm that accomplishes this stage of partitioning is outlined in Table 4 and Fig. 34.

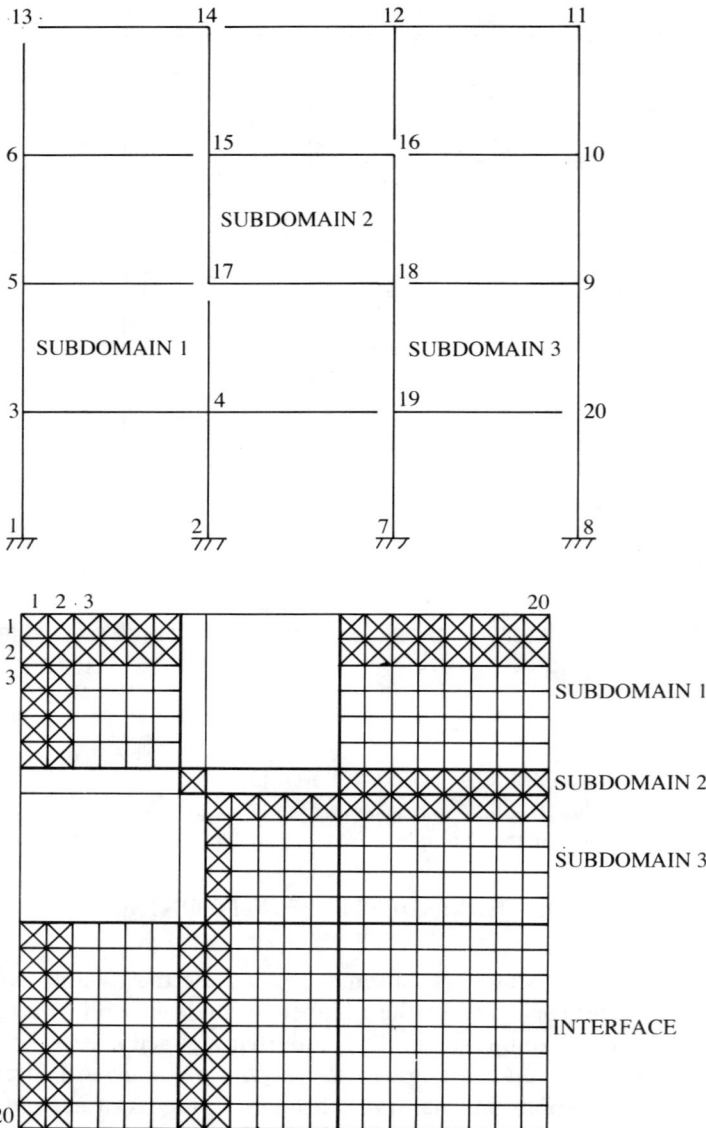

Fig. 33. Renumbering scheme and corresponding stiffness matrix at end of the initial partitioning stage for the example frame of Fig. 25 (case of 3 processors).

TABLE 4
Algorithm for intermediate partitioning

1. Set the node counter to 1. Node 1 is the current active node.
2. If the active node is an internal node, a support, or missing more than one element to be closed, go to step 3. Otherwise, check whether the missing element that closes the interface node opens another internal node. If yes, go to step 3. Otherwise, switch the interface node to an internal one and perform the necessary adjustments in the arrays.
3. If the node counter equals *num_nodes,* the intermedite partitioning stage is complete; stop. Otherwise, increase the node counter by one. This is the current active node; go to step 2.

In essence, this stage attempts to furnish a better starting point for the final partitioning stage: a point with a larger number of internal nodes to be balanced among the processors and a fewer number of interface nodes. This is done in a manner that causes a minimum disturbance (if any) to the features achieved in the initial partitioning stage, so as to maintain at least the same reasonable initial starting

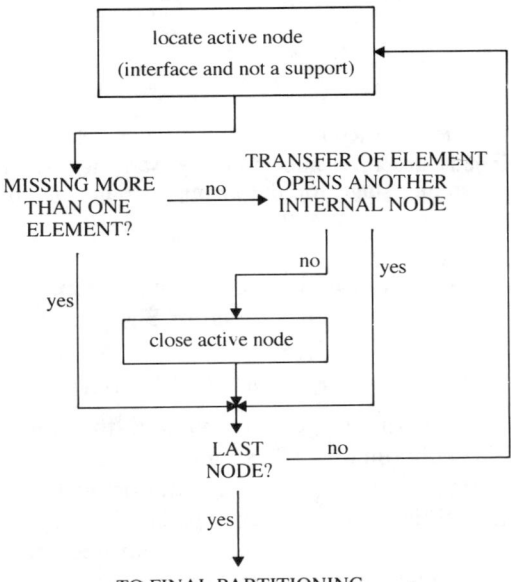

Fig. 34. Flow chart for intermediate partitioning.

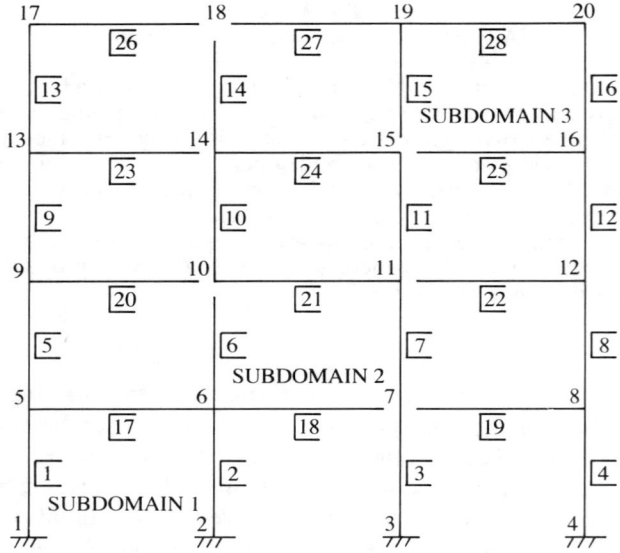

SUBDOMAIN	NUMBER OF INTERNAL NODES	NUMBER OF ELEMENTS
1	5	11
2	0	7
3	5	10

NUMBER OF INTERFACE NODES = 6

Fig. 35. Partitioned structure and domains statistics at the end of the intermediate partitioning stage for the example frame of Fig. 25 (case of 3 processors).

point. The resulting partitioned structure and subdomains statistics are shown in Fig. 35 for the example frame of Fig. 25.

3.6 FINAL PARTITIONING

The main purpose of this stage is to balance the number of internal nodes within the subdomains. This step is crucial, since it attains a load balance among the processors for a major portion of the analysis and design calculations. The key concept is to sweep the subdomains several times, each time with a designated strategy for achieving the desirable balance. The algorithm for final partitioning is outlined in Table 5 and Fig 36.

TABLE 5
Algorithm for final partitioning

1. Set the sweep counter to 1. This is the active sweep.
2. Divide the total number of internal nodes by the number of the processors. This is the 'fair' share of each subdomain. If the sweep counter is less than or equal to 2, set the strategy identifier *strat* to 1; go to step 3. Otherwise, adjust the 'fair' share by increasing it by 1. If any of the last $NP/2$ subdomains possesses an excess of internal nodes over its 'fair' share, set *strat* to 2; go to step 3. If any of the first $NP/2$ subdomains possesses an excess of internal nodes than its 'fair' share, set *strat* to 3; go to step 3. Otherwise, announce the last sweep by setting *strat* to 0; the final partitioning stage is complete; stop.
3. If *strat* = 1 or *strat* = 3, set the subdomain counter to 1. Otherwise, set the subdomain counter to NP. This is the active subdomain.
4. If the subdomain counter exceeds the limits (greater than NP for *strat* = 1 or *strat* = 3, or less than 1 for *strat* = 2), increase the sweep counter by 1; go to step 2. Otherwise, if the active subdomain possesses a number of internal nodes greater than or equal its 'fair' share for *strat* = 1, increase the subdomain counter by 1; repeat the same step 4. If the next higher subdomain possesses a number of internal nodes less than its 'fair' share for *strat* = 2, decrease the subdomain counter by 1; repeat the same step 4. Otherwise, if the next lower subdomain possesses a number of internal nodes less than its 'fair' share for *strat* = 3, increase the subdomain counter by 1; repeat the same step 4. Otherwise, continue.
5. For the active subdomain, scan its interface nodes to determine the estimate-of-damage array, called *mess*. The interface nodes in the *mess* array are labelled according to the following criteria:
   (a) If closing the interface node causes no other internal node in an adjacent subdomain to be opened, label this interface node as 0.
   (b) If closing the interface node causes exactly one internal node in an adjacent subdomain to be opened, label this interface node as 1.
   (c) If closing the interface node causes more than one internal node in one or more adjacent subdomains to be opened, label this interface node as 2.
6. Set the iteration counter to 1. This is the active iteration.
7. If the iteration counter exceeds 3, increase the subdomain counter by 1 for the case of *strat* = 1 or *strat* = 3, or decrease it by 1 for the case of *strat* = 2; go to 4. Otherwise, scan the *mess* array in sequence and pause only at those interface nodes with a label equal to the iteration counter to make necessary transactions of elements and the corresponding adjustments in the partitioning arrays. Increase the iteration counter by 1; repeat the same step 7.

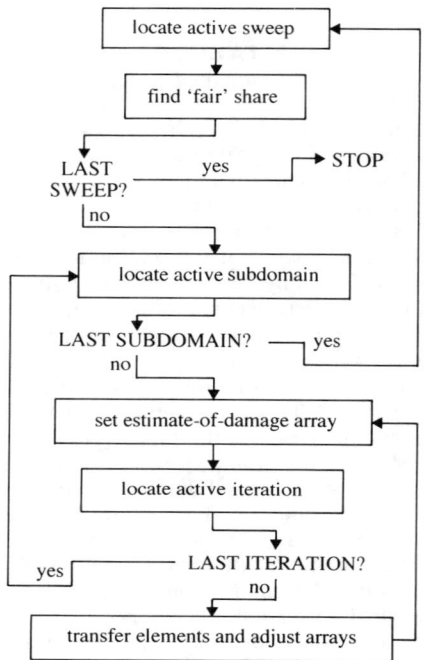

Fig. 36. Flow chart for final partitioning.

The number of sweeps represents the number of times the final partitioning takes place. In the first sweep ($strat = 1$), the subdomains are scanned forward to allow those subdomains with a number of internal nodes less than their 'fair' share to solicit elements from adjacent subdomains with a number of internal nodes greater than their 'fair' share. However, the soliciting process does not terminate when the soliciting subdomain meets its 'fair' share. Rather, the transaction of elements will continue as long as any of the adjacent subdomains has more than its 'fair' share. This helps propagate the excess of internal nodes through the structure from one locality to the other, thus providing remote subdomains with a deficiency in the internal nodes with a better chance of being balanced. The second sweep ($strat = 1$) is a repetition of the first one, but to allow the process of exchange of elements to stabilize.

Much of the node balance will be attained normally in the first sweep and stabilized in the second. However, to account for certain irregularities and for the cases when a subdomain with a shortage of internal nodes is prevented from soliciting from another one with a surplus because of a subdomain with a 'fair' number of internal nodes falling between them, other additional sweeps may be required. When any of the last $NP/2$ subdomains possesses a number of internal nodes more than its 'fair' share in the subsequent sweeps, the subdomains are scanned backward to release some of the internal node accumulation (if any) in the last few subdomains. This strategy ($strat = 2$) is such that a subdomain with an adjacent subdomain higher in order and with a surplus in internal nodes would solicit elements from that neighbour until all its surplus is swept. Otherwise, a similar strategy ($strat = 3$) is adopted when any of the first $NP/2$ subdomains has a surplus of internal nodes, with the exception that the subdomains are scanned forward to undo any accumulation of internal nodes (if any) in the first few subdomains that might have resulted during previous sweeps. The final partitioning stage automatically terminates when none of the subdomains has a surplus of internal nodes.

Generally, the major adjustments are accomplished in sweeps 1 and then 3 (if any), while the remaining sweeps are for refinement. And in all of the sweeps, transactions occur only at those subdomains that meet the previously described criterion or strategy. The essence of this approach is to adjust as much as possible, since the optimal scheme of domain partitioning is not unique in many cases. Figure 37 shows the partitioned structure and subdomain statistics for the final partitioning stage, for the example frame of Fig. 25.

As it turns out, balancing the internal nodes among the subdomains offsets the balance of elements. This matter should be of no concern when running the application program on multiprocessors with shared memory such as the Encore Multimax. In such computers, the shared memory is accessible by all the processors at all times. Therefore, a certain discretization pattern that is suitable for a certain step of the solution process does not have to be dictated to the other steps. Each step can be accommodated with its own discretization pattern or scheme. As an example, the balance of elements achieved in the initial partitioning stage (Fig. 32) can be used for assembling the structure stiffness matrix and evaluation of the element forces and stresses (steps 1 and 5 in Section 3.1, respectively) while the balance of internal nodes attained in the final subdomain stage (Fig. 37) can be used for maintaining better concurrency during solving the linear equations of the entire structure (step 4 in Section 3.1).

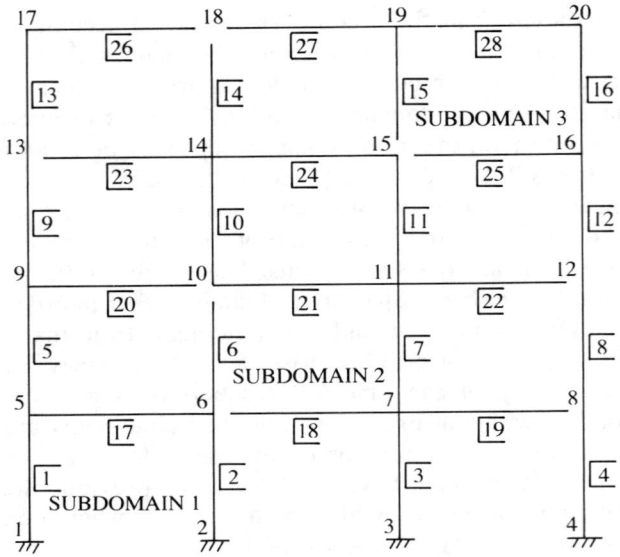

SUBDOMAIN	NUMBER OF INTERNAL NODES	NUMBER OF ELEMENTS
1	3	9
2	3	11
3	3	8

NUMBER OF INTERFACE NODES = 7

Fig. 37. Partitioned structure and domains statistics at the end of the final partitioning stage for the example frame of Fig. 25 (case of 3 processors).

On the other hand, this might seem a problem for multiprocessors with no shared memory, such as the Intel hypercube, since the information for each subdomain is stored in the corresponding processor's local memory, and, therefore, not accessed by any other processors. Consequently, once a discretization pattern has been chosen, it is much less taxing to dictate it for the whole solution process. In this case, the choice is between a balance of internal nodes and a balance of elements. For large structural analysis/optimization problems, the amount of computations involved in the portion that requires a balance of internal nodes (solution of linear equations) is much larger than that portion that requires a balance of elements (assembling the stiffness matrix and evaluation of forces and stresses).

In other words, it is more rewarding in terms of efficiency of the concurrent application to adopt the former balance strategy over the latter. Therefore, the discretization pattern that encompasses a balance of internal nodes overrides the one with a balance of elements, and, therefore, should be used throughout the parallel application for this type of multiprocessor computers. In this sense, the algorithm presented in this book is portable and flexible enough to be used for structural analysis and optimization problems under different multiprocessor computer architectures, and, hence, is machine- or hardware-independent.

3.7 POST-PARTITIONING

The purpose of this stage is to accomplish the following three tasks:

1. Renumber the nodes so as to develop the required decoupling between the subdomain matrices.
2. Decrease the bandwidth within each subdomain and across the interface boundary in order to acquire a better storage scheme and computational efficiency.
3. Prepare for the upcoming analysis and design steps.

In order to renumber the nodes, we need to define two more vectors:

1. *jstart*[$NP + 1$]: a vector that determines the number of internal nodes and supports that are internal within each subdomain. The *jstart* vector for the example frame for the case of three processors is given in Table 6. For example, the number of internal nodes and supports that are internal in subdomain $q(SD_q)$ is equal to the difference between the two numbers in locations q and $q - 1$ of the *jstart* vector. Moreover, the starting number for the new numbering of the internal nodes and supports that are internal in subdomain q is the number stored at the $(q - 1)$th location of the *jstart* array plus 1. Also, the starting number for the new numbering of the interface nodes and supports that are on the interface is the number stored in the location NP of the *jstart* array plus 1.
2. *new_numbering*[*num_nodes*]: a vector that stores the new numbering scheme for the nodes.

TABLE 6
New numbering for the example frame of Fig. 25 (case of 3 processors)

Location	old numbering	new_numbering	jstart
0	1	1	0
1	2	4	5
2	3	9	9
3	4	10	13
4	5	2	
5	6	14	
6	7	6	
7	8	17	
8	9	3	
9	10	16	
10	11	8	
11	12	20	
12	13	15	
13	14	7	
14	15	19	
15	16	12	
16	17	5	
17	18	18	
18	19	11	
19	20	13	

We scan the *nodes_active* array (the array which stores the nodes in the order they are stacked in the initial partitioning stage). We start numbering the nodes according to the subdomain (or interface) to which they belong (the *nodes_trace* array is helpful in this respect). The *new_numbering* vector for the example frame of Fig. 25 is also shown in Table 6. The final pattern of numbering for the example problem and the corresponding stiffness matrix are shown in Fig. 38. Clearly, a better load balance case is achieved over the case shown in Fig. 33.

As mentioned earlier, the algorithm used for the initial partitioning stage results in compact subdomains (i.e. subdomains with contiguous elements). Therefore, use of the *nodes_active* array as a base for the renumbering of the nodes results in subdomains with a small bandwidth. Renumbering the interface nodes is ignored in this work because fill-ins occur in the interface matrix due to condensation of the subdomains matrices.

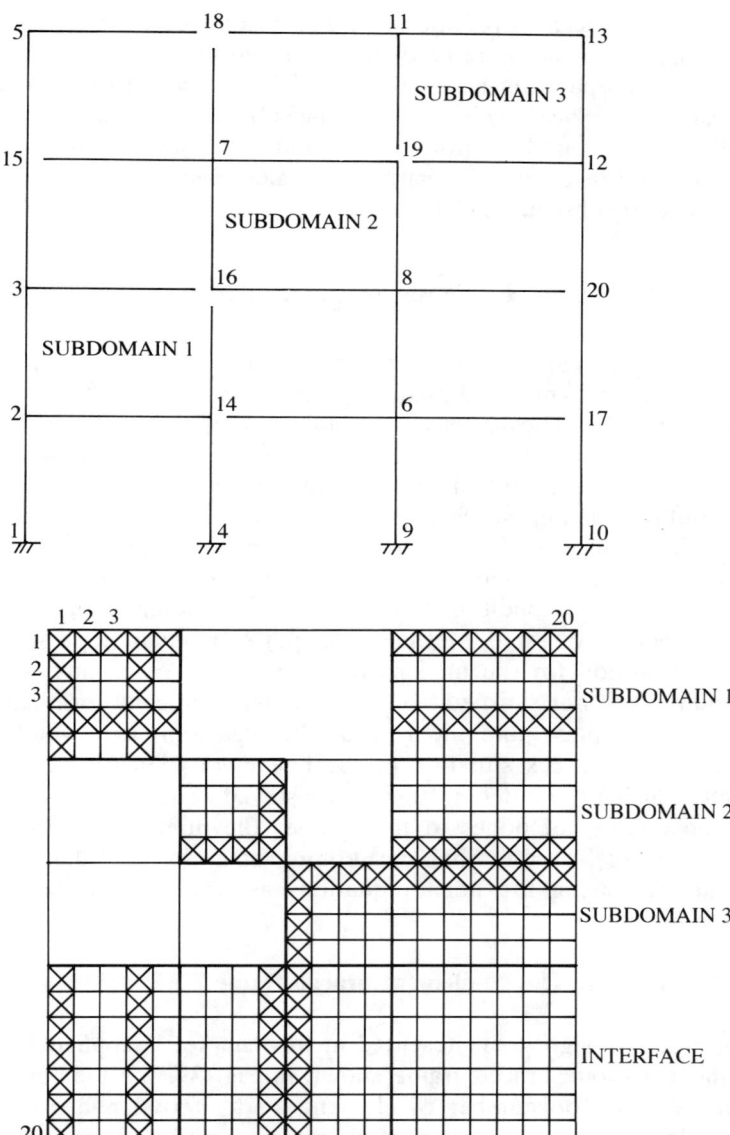

Fig. 38. Renumbering scheme and corresponding stiffness matrix at the end of the final partitioning stage for the example frame of Fig. 25 (case of 3 processors).

Finally, once the final partitioning stage is complete, the remaining analysis and design steps are ready to be carried in parallel. However, some further preparation may be required before concurrent calculations can take place. This is particularly true for multiprocessor computers with shared memory, where different analysis and design steps may require different balancing schemes. Such preparation will be discussed in subsequent chapters.

3.8 APPLICATIONS

The partitioning algorithm presented in this chapter has been applied to several planar frames and planar and space trusses of different size and configuration. The patterns of numbering of the nodes and the elements are shown on their respective figures. Information such as cross-sectional areas, modulus of elasticity, or different loading conditions is not included in this chapter, since the emphasis here is directed towards the balance of elements and internal nodes during each of the partitioning stages. A minimum of nine elements per subdomain is prescribed as a pre-condition for the initial partitioning stage. Similarly, a minimum of three internal nodes per subdomain is prescribed as a pre-condition for the final partitioning stage. When any of the pre-conditions is not satisfied, execution is automatically terminated. In all the examples, note the role of the algorithm as a whole in limiting the bottleneck situation of the linear equation solution step through reducing the number of explicitly solved equations from three times the number of nodes to three times the number of interface nodes. The role of the final partitioning stage is to balance the workload of solving the linear equations associated with the non-interface nodes.

3.8.1 Example 1: The 90-element braced frame

The partitioning algorithm presented in this chapter is applied to the 90-element 47-node braced frame shown in Fig. 39, for the cases of 1–7 processors. The number of elements and internal nodes within each subdomain and the number of the interface nodes for each of the three partitioning stages are summarized in Table 7. In all of the cases, the initial partitioning stage results in a balance of elements, while the internal nodes are not balanced. By the time the final partitioning

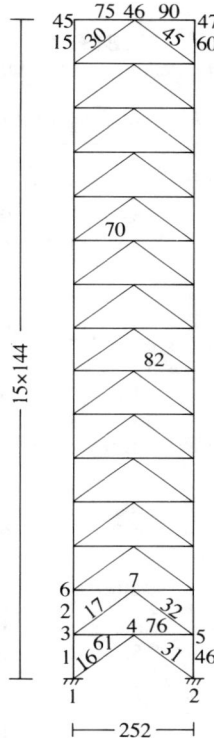

Fig. 39. The 90-element braced frame (all dimensions are in inches).

stage is complete, the balance of nodes is achieved while the balance of elements is no longer maintained. The number of interface nodes is almost the same for the initial and final partitioning stages. Figures 40 and 41 show the partitioning pattern and the subdomains that result for the case of 4 and 7 processors, respectively. In both cases, the compactness of the subdomains is evident.

3.8.2 Example 2: The geodesic dome space truss

The partitioning algorithm is applied to the 132-element 61-node space truss shown in Fig. 42, for the cases of 1–4 processors. The number of elements and internal nodes within each subdomain and the number of the interface nodes for each of the three partitioning stages are

TABLE 7
Summary of partitioning results for the 90-element braced frame

No.	Initial subdomains		Intermediate subdomains		Final subdomains	
	nodes	elmnts	nodes	elmnts	nodes	elmnts
1	45	90	45	90	45	90
	interface = 0		interface = 0		interface = 0	
1	20	45	20	44	20	45
2	21	45	22	46	21	45
	interface = 4		interface = 3		interface = 4	
1	12	30	12	30	12	29
2	11	30	11	30	12	33
3	14	30	14	30	12	28
	interface = 8		interface = 8		interface = 9	
1	9	23	9	23	9	23
2	7	23	8	24	8	24
3	7	22	7	21	9	26
4	10	22	10	22	8	17
	interface = 12		interface = 11		interface = 11	
1	6	18	6	18	6	18
2	5	18	5	18	5	18
3	5	18	5	18	5	17
4	5	18	5	18	6	21
5	8	18	8	18	6	16
	interface = 16		interface = 16		interface = 17	
1	5	15	5	14	5	14
2	3	15	4	16	4	16
3	4	15	4	14	4	14
4	3	15	4	16	4	15
5	4	15	4	14	4	17
6	6	15	7	16	5	14
	interface = 20		interface = 17		interface = 19	
1	4	13	5	14	4	13
2	3	13	3	12	4	15
3	3	13	3	12	3	10
4	2	13	4	15	3	14
5	2	13	2	12	3	12
6	2	13	2	13	3	16
7	5	13	5	12	3	10
	interface = 24		interface = 21		interface = 22	

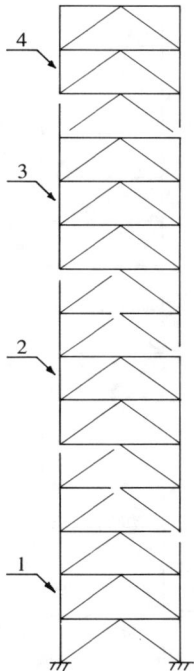

Fig. 40. Partitioning of the 90-element braced frame (case of 4 processors).

summarized in Table 8. In all of the cases, the balance of elements attained in the initial partitioning stage is offset on the account of balancing the number of internal nodes in the final partitioning stage. As an example, the initial partitioning stage for the case of 4 processors results in 33 elements within each subdomain, 21 interface nodes, and a maximum difference of 5 internal nodes (15 linear equations) between the first and second subdomains. Once the final partitioning stage is complete, the difference between the subdomains 2 and 4 is 17 elements, the number of interface nodes is 22, and a balance of internal nodes is achieved. Figures 43 and 44 show the partitioning pattern and the subdomains that result for the case of 3 and 4 processors, respectively.

3.8.3 Example 3: The 200-bar plane truss

The partitioning algorithm is also applied to the 200-element 77-node plane truss shown in Fig. 45, for the cases of 1, 2, 4, 6, and 8

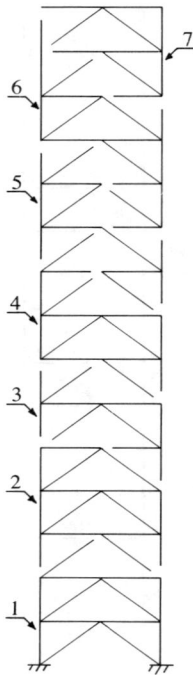

Fig. 41. Partitioning of the 90-element braced frame (case of 7 processors).

processors. The number of elements and internal nodes within each subdomain and the number of the interface nodes for each of the three partitioning stages are summarized in Table 9. Note the balances attained in the initial and final partitioning stages. For example, the initial partitioning stage for the case of 8 processors results in 25 elements within each subdomain, 60 interface nodes, and a difference of 8 internal nodes (24 linear equations) between the first and third subdomains. After the final partitioning stage has been completed, the difference between subdomains 1 and 2 is 33 elements, the number of interface nodes is reduced to 40 (a reduction of 60 interface linear equations) and a balance of internal nodes is achieved. The role of the intermediate partitioning stage in reducing the number of interface nodes and consequently increasing the number of internal nodes is evident in this case. Figures 46 and 47 show the partioning pattern and

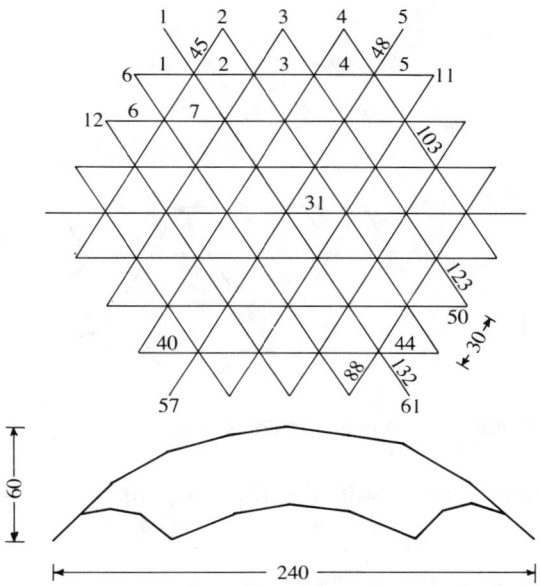

Fig. 42. Geodesic dome space truss (all dimensions are in inches).

TABLE 8
Summary of partitioning results for the geodesic dome space truss

No.	Initial subdomains		Intermediate subdomains		Final subdomains	
	nodes	elmnts	nodes	elmnts	nodes	elmnts
1	37	132	37	132	37	132
	interface = 0		interface = 0		interface = 0	
1	16	66	16	66	15	65
2	14	66	14	66	15	67
	interface = 7		interface = 7		interface = 7	
1	10	44	10	44	7	41
2	6	44	6	44	8	54
3	7	44	7	44	7	37
	interface = 14		interface = 14		interface = 15	
1	7	33	7	33	4	30
2	2	33	2	33	4	45
3	3	33	3	33	3	29
4	4	33	4	33	4	28
	interface = 21		interface = 21		interface = 22	

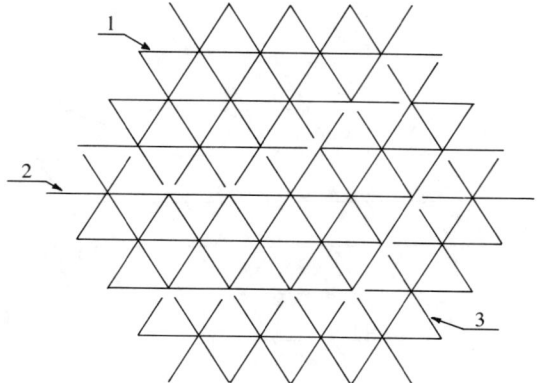

Fig. 43. Partitioning of the geodesic dome space truss (case of 3 processors).

the subdomains that result for the case of 4 and 8 processors, respectively.

3.8.4. Example 4: The 266-element frame

The partitioning algorithm is applied to the 266-element 153-node plane frame shown in Fig. 48, for the cases of 1, 2, 4, 6, 8, and 10 processors. This example is chosen to demonstrate the effectiveness of the algorithm for structures of irregular configurations such as frames

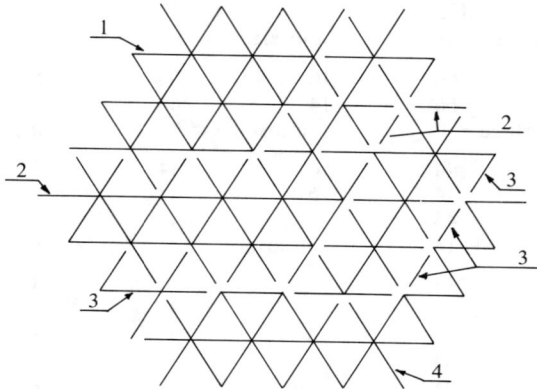

Fig. 44. Partitioning of the geodesic dome space truss (case of 4 processors).

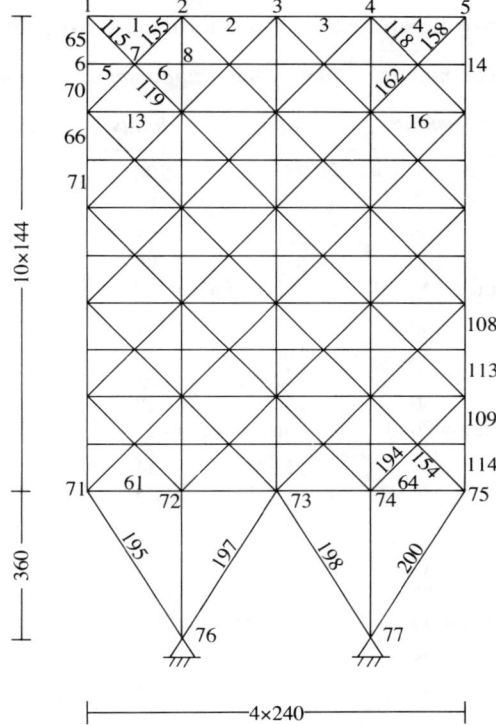

Fig. 45. 200-bar plane truss (all dimensions are in inches).

with setbacks. The number of elements and internal nodes within each subdomain and the number of the interface nodes for each of the three partitioning stages are summarized in Table 10. A comparison between the number of the interface nodes in the initial and final partitioning stages shows that it is the same for the case of 2 processors, while it increases for the cases of 4 and 6 processors and decreases for the cases of 8 and 10 processors. For example, the initial partitioning stage for the case of 10 processors results in 26 or 27 elements within each subdomain, 77 interface nodes, and a maximum difference of 8 internal nodes (24 linear equations) between the first and fourth subdomains. After the final partitioning stage has been completed, there is a difference of 16 elements between subdomains 8 and 10, the number of interface nodes is reduced to 61 (a reduction of

TABLE 9
Summary of partitioning results for the 200-bar plane truss

No.	Initial subdomains		Intermediate subdomains		Final subdomains	
	nodes	elmnts	nodes	elmnts	nodes	elmnts
1	75	200	75	200	75	200
	interface = 0		interface = 0		interface = 0	
1	35	100	36	101	33	98
2	31	100	31	99	33	102
	interface = 9		interface = 8		interface = 9	
1	16	50	17	51	13	47
2	9	50	10	50	13	6
3	10	50	12	51	12	45
4	14	50	14	48	13	47
	interface = 26		interface = 22		interface = 24	
1	11	34	11	34	7	27
2	3	34	4	35	7	45
3	4	33	6	32	6	35
4	4	33	6	32	6	22
5	3	33	4	33	6	38
6	6	33	7	34	6	33
	interface = 44		interface = 37		interface = 37	
1	8	25	8	25	5	15
2	1	25	2	23	5	48
3	0	25	4	24	4	15
4	0	25	5	27	4	17
5	0	25	3	23	4	30
6	1	25	7	30	4	22
7	2	25	3	21	5	29
8	3	25	6	27	4	24
	interface = 60		interface = 37		interface = 40	

48 interface linear equations) and a balance of internal nodes is achieved. Figures 49 and 50 show the partitioning pattern and the subdomains that result for the case of 6 and 10 processors, respectively.

3.9 SUMMARY AND CONCLUSIONS

An efficient three-stage algorithm for partitioning of frame structures (made of two-node elements) has been presented. The input data

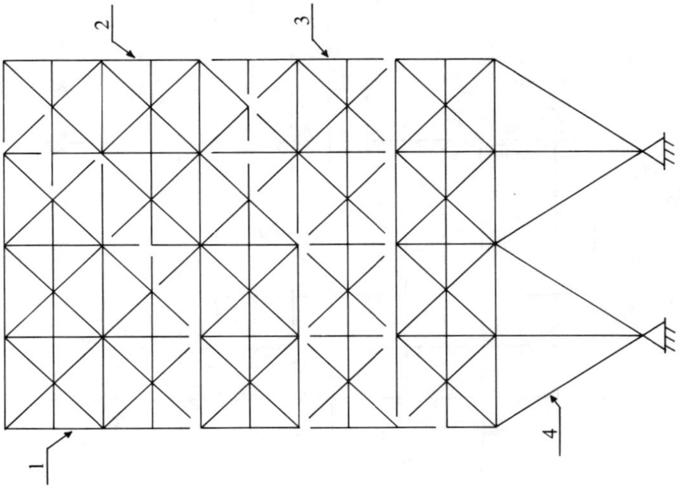

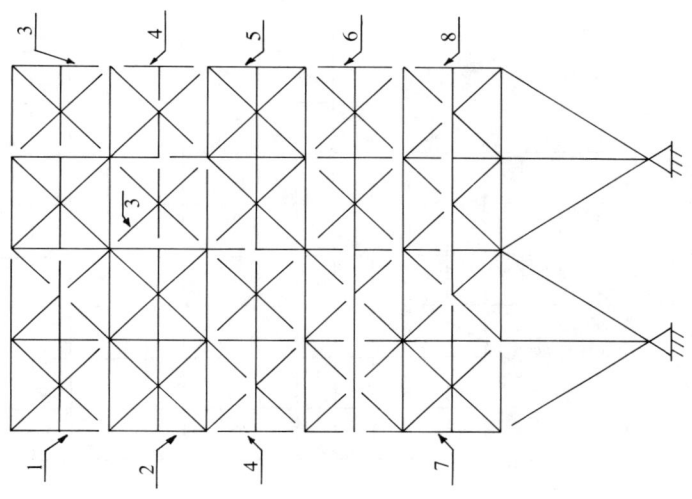

Fig. 46. Partitioning of the 200-bar plane truss (case of 4 processors).

Fig. 47. Partitioning of the 200-bar plane truss (case of 8 processors).

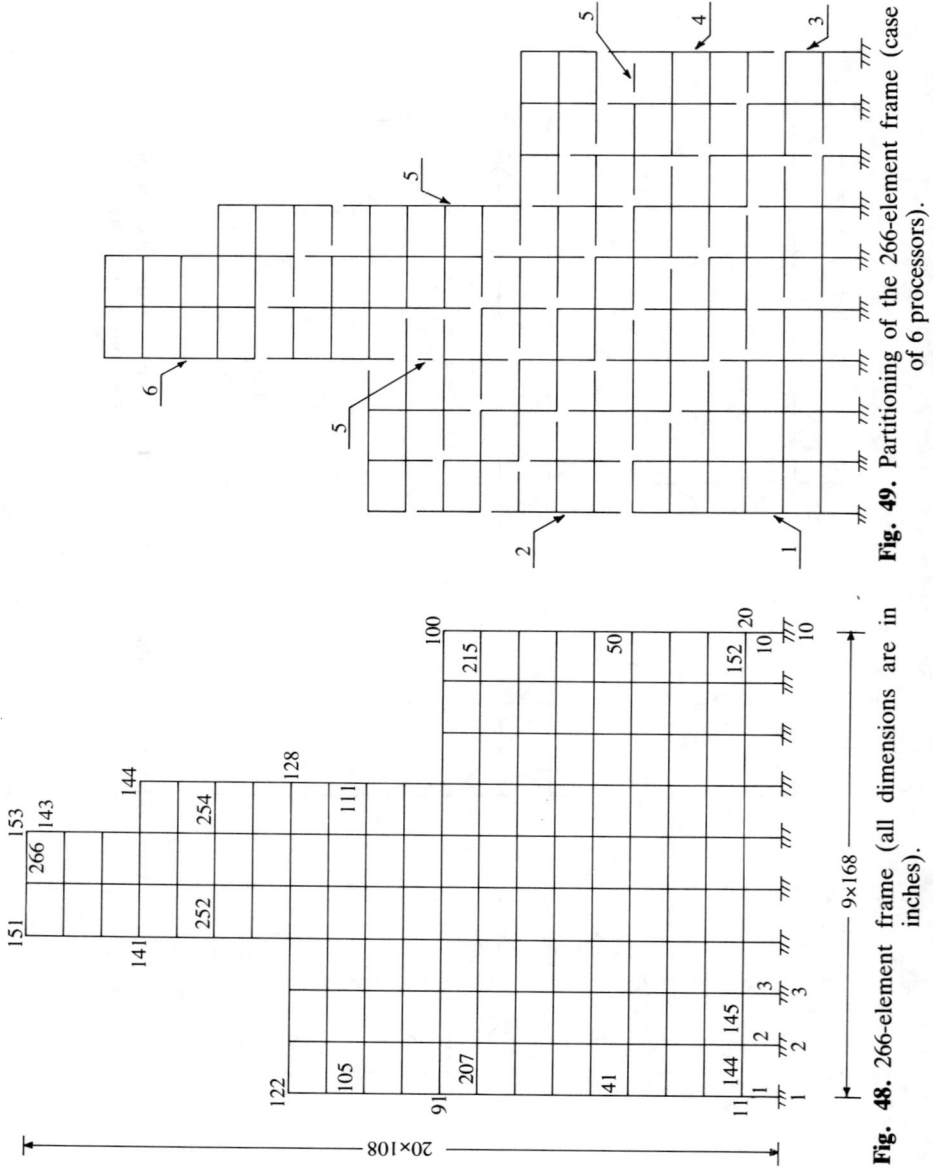

Fig. 49. Partitioning of the 266-element frame (case of 6 processors).

Fig. 48. 266-element frame (all dimensions are in inches).

TABLE 10
Summary of partitioning results for the 266-element frame

No.	Initial subdomains		Intermediate subdomains		Final subdomains	
	nodes	elmnts	nodes	elmnts	nodes	elmnts
1	143	266	143	266	143	266
	interface = 0		interface = 0		interface = 0	
1	62	133	62	133	66	140
2	70	133	70	133	66	126
	interface = 11		interface = 11		interface = 11	
1	30	67	30	67	27	62
2	23	67	23	67	26	78
3	26	66	26	66	28	74
4	36	66	36	66	28	52
	interface = 28		interface = 28		interface = 34	
1	20	45	20	45	16	41
2	14	45	14	45	16	47
3	11	44	11	44	12	45
4	14	44	14	44	16	54
5	16	44	17	45	16	48
6	24	44	24	43	16	31
	interface = 44		interface = 43		interface = 51	
1	14	34	14	34	11	31
2	10	34	10	34	11	36
3	7	33	7	33	10	35
4	5	33	5	32	9	34
5	7	33	8	34	10	34
6	9	33	9	33	11	40
7	13	33	13	32	11	33
8	18	33	19	34	11	23
	interface = 60		interface = 58		interface = 59	
1	11	27	11	27	8	24
2	7	27	7	27	8	32
3	4	27	4	26	8	30
4	3	27	4	27	8	26
5	3	27	4	27	7	23
6	4	27	7	30	8	22
7	4	26	4	24	8	29
8	6	26	6	26	9	34
9	10	26	10	25	9	28
10	14	26	15	27	9	18
	interface = 77		interface = 71		interface = 61	

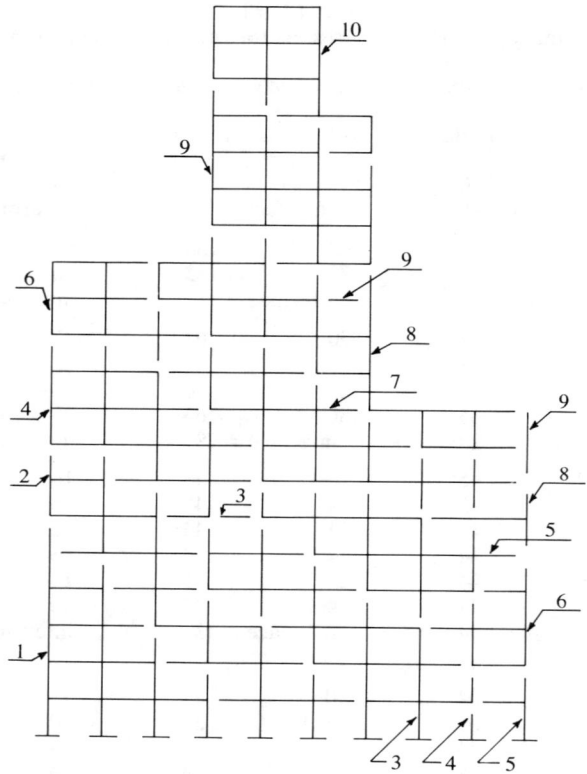

Fig. 50. Partitioning of the 266-element frame (case of 10 processors).

(information about nodes and elements) is pre-processed in a manner that enables subsequent analysis and design steps to be carried out efficiently on multiprocessor computers. The algorithm achieves concurrency on the global level (the structure level), with the ultimate objective of balancing the number of internal nodes within the subdomains while attaining a small number of interface nodes. This strategy of load balance of internal nodes can be generalized for all of the analysis and design steps of the solution process for applications on multiprocessors with no shared memory. This results from the fact that balancing the number of internal nodes is crucial for a major step of the analysis process (i.e. the solution of linear equations).

For multiprocessors with shared memory, this strategy can be used for solution of the linear equations step only, while other steps can be supplied with their most suitable load balancing schemes. In this case, we are back to concurrency on the local level (the steps level). In this sense, the algorithm presented here is portable and flexible enough to be used for structural analysis and design problems under different multiprocessor computer architectures, and, hence, is machine- or hardware-independent.

The generation of the initial partitioning and all subsequent adjustments during the intermediate and final partitioning stages are based on connectivity, thus making the algorithm quite general in the sense that it can handle irregularities in configuration or numbering schemes. This connectivity strategy, in addition to the intermediate partitioning stage, helps reduce the number of interface nodes, and, hence, the bottleneck situation that results during the solution of the interface linear equations. Also, all the vectors involved in the substructuring algorithm are integers, requiring a small storage and resulting in calculations that can be processed rapidly. This makes the algorithm especially suitable for the analysis and design of large structures.

Chapter IV

Concurrent Analysis of Structures

4.1 INTRODUCTION

A number of researchers have presented algorithms for concurrent processing of structures modeled by finite elements during the past few years (Farhat & Wilson [1987]; Farhat *et al.* [1987*b*]; Lou & Friedman [1989]; Noor & Peters [1989]). In this chapter, we present algorithms for concurrent analysis of framed structures consisting of two-node elements such as plane frames and space trusses.

In Chapter III, we presented a three-stage algorithm for automatic partitioning of framed structures suitable for concurrent processing (Kamal & Adeli [1990, 1991]). In this chapter, we extend that work by presenting algorithms and procedures for the complete analysis of structures on multiprocessor computers. The partitioning of the structure is suceeded by seven steps using the stiffness (displacement) method of structural analysis (Fig. 51). These steps are as follows: assembling the structure stiffness matrix, setting up the load vector, applying the boundary conditions, condensation of the non-interface displacement degrees of freedom onto the interface ones, solution of the interface displacement degrees of freedom, retrieving the non-interface displacement degrees of freedom, and computation of the element forces and stresses.

The rest of the chapter is organized as follows. The equilibrium equations resulting form substructuring algorithm are briefly described in Section 4.2. The assembly of the structure stiffness matrix and the setting up of the load vector are given in Sections 4.3 and 4.4. The application of the boundary conditions is discussed in Section 4.5. The

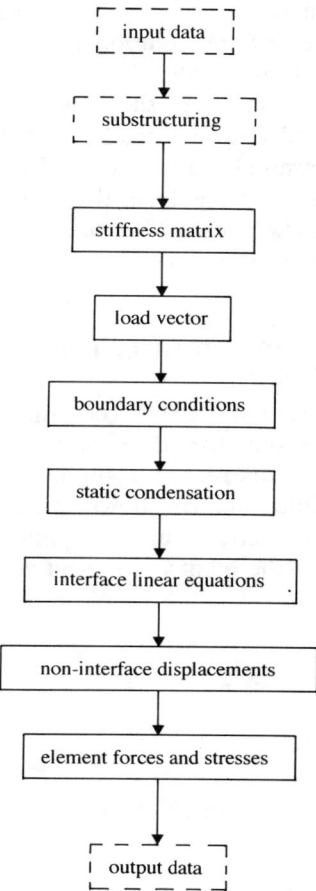

Fig. 51. Main steps of analysis of structures by the displacement (stiffness) method.

static condensation of the subdomains and the solution of the interface linear equations are described in Sections 4.6 and 4.7. Then, computation of the generalized non-interface displacements is presented in Section 4.8, followed by the evaluation of the element forces and stresses in Section 4.9. The application of the parallel algorithms presented in this chapter to several truss and frame problems is given in Section 4.10. The chapter ends with conclusions in Section 4.11.

It should be noted how concurrency is achieved and implemented in this work. In each step of the solution process mentioned earlier, a certain number of sets is automatically generated by the program. This number is equal to the number of the prescribed processors. Each set is automatically mapped onto a thread. Thus, the number of threads created is equal to the number of prescribed processors. This strategy is chosen on the basis of results from the numerical experimentations presented in Chapter II. Encore Multimax devices dynamically schedule the threads to the processors.

4.2 EQUILIBRIUM EQUATIONS

In Chapter III, we presented the substructuring algorithm. Application of the algorithm to the four-story three-bay frame shown in Fig. 25 for the case of three processors results in the new numbering scheme and the corresponding stiffness matrix shown in Fig. 52 (at the end of the final partitioning stage). The resulting equilibrium equations can be summarized as follows (the blanks represent zero terms):

$$\mathbf{K}_{II}(1)\mathbf{r}_I(1) \qquad\qquad\qquad + \mathbf{K}_{IB}(1)\mathbf{r}_B = \mathbf{R}_I(1) \tag{7}$$

$$\mathbf{K}_{II}(2)\mathbf{r}_I(2) \qquad\qquad + \mathbf{K}_{IB}(2)\mathbf{r}_B = \mathbf{R}_I(2) \tag{8}$$

$$\ddots \qquad\qquad\qquad \vdots \qquad\quad \vdots$$

$$\mathbf{K}_{II}(q)\mathbf{r}_I(q) \quad + \mathbf{K}_{IB}(q)\mathbf{r}_B = \mathbf{R}_I(q) \tag{9}$$

$$\vdots \qquad\qquad \vdots$$

$$\mathbf{K}_{BI}(1)\mathbf{r}_I(1) + \mathbf{K}_{BI}(2)\mathbf{r}_I(2) \ldots + \mathbf{K}_{BI}(q)\mathbf{r}_I(q) \ldots + \mathbf{K}_{BB}\mathbf{r}_B = \mathbf{R}_B \tag{10}$$

The strong decoupling of the diagonal blocks $\mathbf{K}_{II}(q)$ suggests that they can be processed concurrently and their effect can be collapsed onto the interface matrix $\mathbf{K}_{BB}$. Therefore, in this approach, the only set of linear equations that need be solved simultaneously is the one associated with the interface. This set is generally much smaller than the original one. This helps alleviate much of the slow-down in the speed-up that occurs during the inherently sequential process of the solution of the linear simultaneous equations.

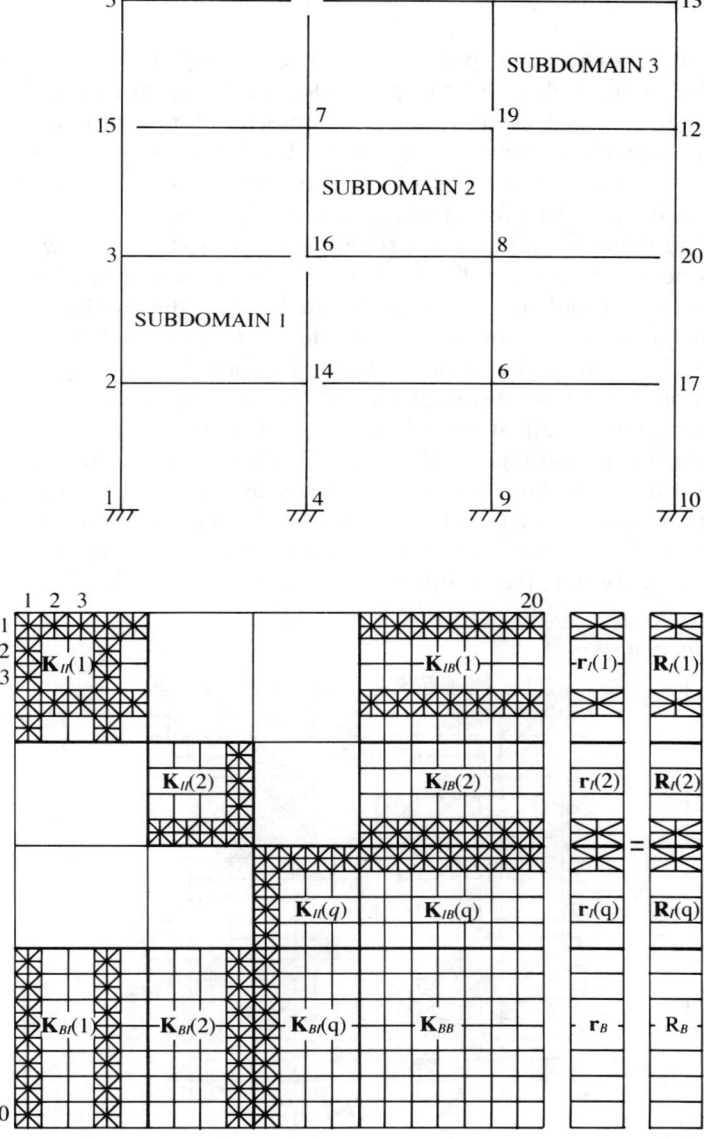

Fig. 52. Numbering scheme and the corresponding stiffness matrix at the end of the final partitioning stage for a 3-bay 4-storey frame (case of 3 processors).

4.3 ASSEMBLING THE STRUCTURE STIFFNESS MATRIX

Once the partitioning process has been completed, the resulting stiffness matrix will be of the decoupled form shown in Fig. 53. The cross-hatched squares represent 3×3 blocks of nonzero terms. The non-cross-hatched squares represent 3×3 blocks of zero terms. Obviously, there is a large number of zero terms in the structure stiffness matrix. To avoid storing such zero terms, the domain of the structure stiffness matrix is classified into four zones as shown in Fig. 53. Zone 1 encompasses the degrees of freedom corresponding to the internal nodes and support nodes located within the subdomains. The skyline (or active column) storage scheme (Bathe [1982]) is used in this work to store the nonzero terms in zone 1 in a single column shown on the left-hand side of Fig. 53. Because of symmetry, only the upper-triangular half of zone 1 is stored. The envelope of the skyline for this zone is also identified in Fig. 53. Occasionally, some blocks of zero terms, like those in rows 7 and 8 and column 9, are stored. Fill-ins, however, occur in these locations during the subsequent steps of the solution process. The blocks in zone 2 and zone 3 represent the coupling between the subdomains and the interface. Because of

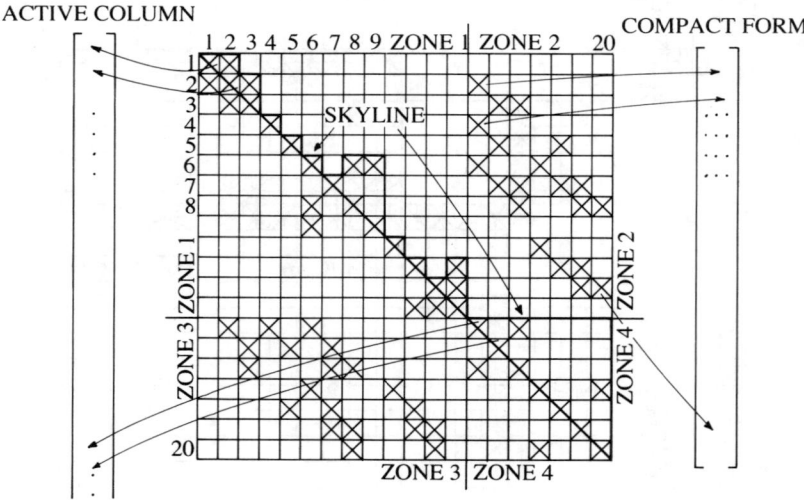

Fig. 53. Different zones and storing schemes for the structure stiffness matrix for the example frame of Fig. 52 for the case of 3 processors.

symmetry, only the nonzero blocks in zone 2 are stored in a compact three-column form shown on the right-hand side of Fig. 53. Finally, zone 4 represents the degrees of freedom associated with the interface nodes and supports on the interface. Since fill-ins occur during the solution process, and because of symmetry, only the upper half of the interface matrix is stored and operated on. As shown in Fig. 53, this portion of the structure stiffness matrix is also stored in the lower portion of the skyline array.

A balanced number of elements can now be mapped to each thread. Encore Multimax devices assign the threads to the processors. The processors concurrently evaluate the element stiffness matrices and then assemble them into the structure stiffness matrix. When two processors are updating the same location of the structure stiffness matrix simultaneously, a 'racing condition' may occur, and, consequently, incorrect results may be the outcome. A possibly remedy to such a situation is to synchronize the updating of the locations of the structure stiffness matrix susceptible to simultaneous access by more than one processor. Obviously, this synchronization process slows down the speed-up expected for the concurrent application (Chien & Sun [1989]). Consequently, the question becomes that of how a racing condition can be overcome while maintaining a minimum jeopardy to the efficiency of the concurrent application.

In Chapter II, we adopted semaphores and monitors for synchronization. It was concluded that semaphores require less time to be invoked than monitors, and, thus, result in a better concurrent performance. On the other hand, we assigned semaphores only to the locations that were susceptible to a racing condition. The solid boxes in Fig. 54(a) show that the number of such locations is 9 times the *number of nodes* (each solid box represents a 3 × 3 matrix). When the banded storage scheme or the skyline storage scheme is used, this number is reduced to 6 times the *number of nodes*, as shown by the filled triangles in Fig. 54(b). Still, synchronization can slow down the speed-up expected for the concurrent performance of this step significantly. Chien & Sun [1989] suggested that a racing condition can be avoided by renumbering the elements in such a manner that no location of the structure stiffness matrix is addressed simultaneously by more than one processor. However, there are times when this approach will not work, like in the case of many processors with a small number of elements assigned to each one of them.

In this chapter, we investigate two strategies for avoiding a racing

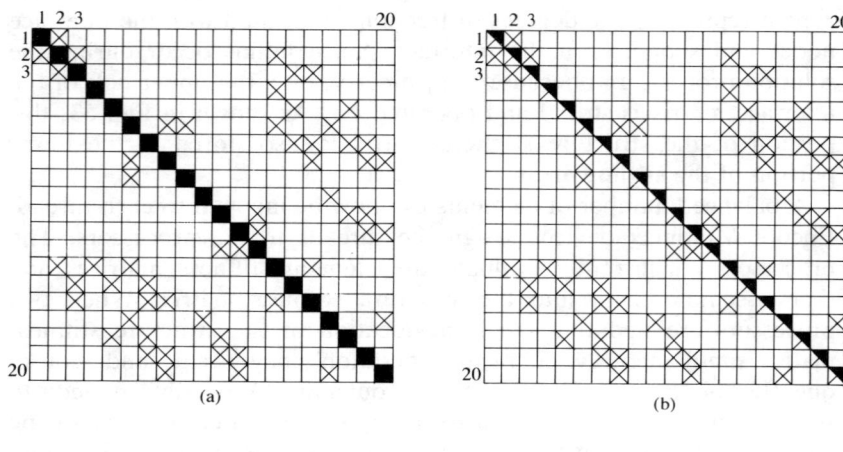

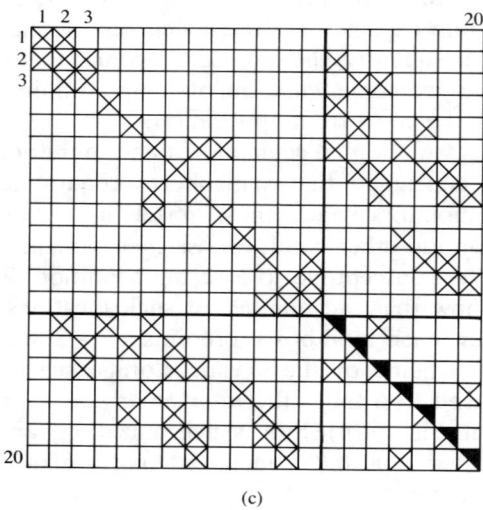

Fig. 54. Illustration of the overlapped locations in the structure stiffness matrix subjected to a racing condition for the example frame of Fig. 52 for the case of three processors, for (a) complete structure stiffness matrix, (b) skyline storage scheme, and (c) interface skyline storage scheme.

condition. In both strategies, we employ the results of the initial partitioning stage. In this stage, the number of elements is balanced among the processors; therefore, a balance of workload amongst them is maintained. On the other hand, in this case, the number of locations requiring synchronization is equal to only 6 times the *number of interface nodes*. This is due to the fact that synchronization is required only when a number of processors are addressing the stiffness matrices of elements with common nodes in the structure stiffness matrix simultaneously. Obviously, such a node has to be an interface node. The filled triangles in Fig. 54(c) give a clear view of the overlapped locations for the example frame of Fig. 52 for the case of three processors.

In the first strategy, we use a number of semaphores equal to the number of overlapped locations (6 times the *number of interface nodes*). Unlike the strategy present in Chapter II, where the number of semaphores was fixed, the number of semaphores in this strategy is equal to the number of interface nodes (which usually increases with the number of subdomains or processors). For example, the overhead time required for invocation of the semaphores in Chapter II is included, even for the case of one processor when no synchronization is actually required. In the present strategy, the case of one processor or subdomain will not have any interface node, and therefore no synchronization is required and no semaphores are invoked. However, the overhead time due to invocation of semaphores tend to be more significant as the number of subdomains and consequently the number of interface nodes increases. This overhead, however, is compensated to some extent by the presence of more processors (Adeli & Kamal [1989*a*, *b*]).

The other strategy is to avoid the racing condition completely through the use of extra storage locations for those overlapped sections (6 times the *number of interface nodes*). After the assembly process has been completed, the addition of the terms in the overlapped locations can take place in parallel through the creation of a new set of threads to accomplish just that.

Table 11 summarizes the algorithms for the two aforementioned strategies. All the sections that need to be synchronized are short in length (only one addition) and cheap to evaluate. We compare the effect of the overhead time required to synchronize the threads in the first strategy with that required to introduce an extra addition step and create a set of threads to parallelize this step in the second strategy.

TABLE 11
Algorithms for setting up the structure stiffness matrix

Strategy (a): Create threads and semaphores
1. Distribute the number of elements as evenly as possible among a number of sets equal to the number of prescribed processors. The balance of elements achieved at the end of the initial partitioning stage is used to determine these sets.
2. Create a number of threads equal to the number of prescribed processors. Each set is mapped onto a thread.
3. For $q = 1, \ldots, num\_threads$, do concurrently
   For $elmnt = 1, \ldots, num\_assigned\_elmnts$, do sequentially
   1. Set up the element stiffness matrix.
   2. If any end node of the element is an interface node, synchronize when assembling the portion of the stiffness matrix corresponding to this node and falling in the overlapped area of zone 4. Otherwise, do not synchronize.
   Next *elmnt*.

Strategy (b): Create threads only
1. Repeat steps 1–3 in strategy (a), with the following exception. Avoid a synchronization situation through using extra storage locations for the overlapped locations of zone 4 of the structure stiffness matrix.
2. Distribute this number of the overlapped locations of the interface portion (zone 4) of the structure stiffness matrix as evenly as possible among a number of sets equal to the number of prescribed processors.
3. Create a number of threads equal to the number of prescribed processors. Each set is mapped onto a thread.
4. For $q = 1, \ldots, num\_threads$, do concurrently
   For $add = 1, \ldots, num\_assigned\_additions$, do sequentially
   1. Add-up the individual components of the overlapped location.
   2. Store the sum in its apppropriate location of the structure stiffness matrix.
   Next *add*.

4.4 LOAD VECTOR

The load vector is composed of two types of loads. The first type is due to initial stresses and loads acting along or at some points of the spans of the elements. The other type is due to loads acting at the nodes. Since the first type is associated with the elements, the portion of the load vector due to these loads is set up during assembling the structure stiffness matrix. Elements with one or both of their nodes falling on the interface require synchronization when simultaneous access to the interface portion of the load vector is encountered.

Strategies similar to those for the case of assembling the structure stiffness matrix (outlined in Table 11) are adopted for assembling the load vector due to this type of loads. The second type of loads is associated with the nodal data. The number of nodes can be evenly mapped into a number of threads equal to the number of prescribed processors. No synchronization is required during assembling the second type of loads into the load vector.

4.5 BOUNDARY CONDITIONS

The next step is to apply the boundary conditions. Basically, we account for the restrained degrees of freedom in this step. This entails adjusting the load vector and the structure stiffness matrix to include this effect. Once this has been done, the restrained degrees of freedom and the corresponding rows in the load vector and rows and columns in the stiffness matrix are ignored in the subsequent solution steps until the displacements are determined. In order to balance the workload in this step, the number of restrained degrees of freedom is evenly mapped onto a number of threads equal to the number of prescribed processors. Simultaneous updating of load vector locations may be encountered, and racing conditions may occur. To overcome this problem, strategies similar to those outlined in Table 11 are used in this step.

4.6 STATIC CONDENSATION

The Gaussian elimination procedure for solution of the set of linear equilibrium equations $\mathbf{Kr} = \mathbf{R}$ for the generalized displacements vector $\mathbf{r}$ consists of three main stages: transforming the matrix into an upper-triangular one, adjusting the load vector, and backward substitution.

In this work, we adopt a variation of this procedure called the Cholesky decomposition approach, in which the amount of arithmetic operations is considerably smaller than the Gaussian elimination (Burden *et al.* [1981]). Noting that the structure stiffness matrix $\mathbf{K}$ is symmetric and positive-definite; it is uniquely decomposed as (Bathe

[1982])

$$K = LDL^T \qquad (11)$$

where L is a lower-triangular matrix with 1s along its diagonal, D is a diagonal matrix, and the superscript T refers to the matrix transpose. This stage, called factorization, corresponds to the first stage in the Gaussian elimination procedure. The product matrix DL^T is the upper-triangular matrix in the Gaussian elimination procedure. The equations used in finding the elements of the matrices L and D for the skyline storage scheme are described later in this section. In order to reduce the storage requirements, the elements of L^T and D are stored in place of the elements of the matrix K.

Once the structure stiffness matrix has been factorized, the equilibrium equations $Kr = R$ can be expressed as

$$(LDL^T)r = R \qquad (12)$$

Defining the vector Q as

$$Q = DL^T r \qquad (13)$$

Equation (12) can be written in the form

$$LQ = R \qquad (14)$$

This equation is solved for the vector Q through forward substitution. This stage is equivalent to adjusting the load vector in the Gaussian elimination procedure, while the vector Q represents the adjusted load vector. Finally, eqn (13) is solved through backward substitution for the generalized displacements vector r. This stage is identical to the backward substitution step in the Gaussian elimination procedure. The equations involved in forward and backward substitutions will be described later in this section.

The above procedure, although presented with respect to the entire structure stiffness matrix, can be applied to the subdomain stiffness matrices. For example, consider a subdomain q. The diagonal matrix $K_{II}(q)$ is factorized into $L_{II}(q)D_{II}(q)L_{II}^T(q)$. The interface decoupling matrix $K_{IB}(q)$ can be written as $L_{II}(q)D_{II}(q)L_{II}^T(q)U_{IB}(q)$, where $U_{IB}(q)$ is obtained through the use of the forward and backward substitution equations. Similarly, $R_I(q)$ can be written as $L_{II}(q)D_{II}(q)L_{II}^T(q)V_I(q)$. Once the same operations have been repeated for all the subdomains, the terms in eqns (7)–(9) are substituted by their corresponding expressions. Normalizing with respect to $\mathbf{L}_{II}(q)\mathbf{D}_{II}(q)\mathbf{L}_{II}^T(q)$, eqns (7)–(9) can now be expressed as (where the blank terms represent zeros):

$$\mathbf{r}_I(1) \qquad\qquad + \mathbf{U}_{IB}(1)\mathbf{r}_B = \mathbf{V}_I(1) \qquad (15)$$
$$\mathbf{r}_I(2) \qquad\qquad + \mathbf{U}_{IB}(2)\mathbf{r}_B = \mathbf{V}_I(2) \qquad (16)$$
$$\ddots \qquad\qquad \vdots \qquad\quad \vdots$$
$$\mathbf{r}_I(q) \qquad\qquad + \mathbf{U}_{IB}(q)\mathbf{r}_B = \mathbf{V}_I(q) \qquad (17)$$
$$\ddots \qquad\qquad \vdots \qquad\quad \vdots$$

Equations such as (15)–(17) suggest that the displacements of internal nodes (within the subdomains) can be expressed in terms of the interface displacements as follows:

$$\mathbf{r}_I(1) = \mathbf{V}_I(1) - \mathbf{U}_{IB}\mathbf{r}_B \qquad (18)$$
$$\mathbf{r}_I(2) = \mathbf{V}_I(2) - \mathbf{U}_{IB}(2)\mathbf{r}_B \qquad (19)$$
$$\vdots \qquad \vdots \qquad \vdots$$
$$\mathbf{r}_I(q) = \mathbf{V}_I(q) - \mathbf{U}_{IB}(q)\mathbf{r}_B \qquad (20)$$
$$\vdots \qquad \vdots \qquad \vdots$$

Substituting the expressions for the internal (non-interface) displacements obtained from equations such as (18)–(20) into the interface displacments (equilibrium) equation defined by eqn (10) and rearranging terms, we obtain the following equation:

$$[\mathbf{K}_{BB} - \mathbf{K}_{BI}(1)\mathbf{U}_{IB}(1) - \mathbf{K}_{BI}(2)\mathbf{U}_{IB}(2) - \ldots - \mathbf{K}_{BI}(q)\mathbf{U}_{IB}(q) \ldots]\mathbf{r}_B$$
$$= \mathbf{R}_B - \mathbf{K}_{BI}(1)\mathbf{V}_I(1) - \mathbf{K}_{BI}(2)\mathbf{V}_I(2) - \ldots - \mathbf{K}_{BI}(q)\mathbf{V}_I(q) \ldots \qquad (21)$$

or

$$\mathbf{K}_{BB}^*\mathbf{r}_B = \mathbf{R}_B^* \qquad (22)$$

where

$$\mathbf{K}_{BB}^* = \mathbf{K}_{BB} - \mathbf{K}_{BI}(1)\mathbf{U}_{IB}(1) - \mathbf{K}_{BI}(2)\mathbf{U}_{IB}(2) - \ldots - \mathbf{K}_{BI}(q)\mathbf{U}_{IB}(q) \ldots$$
$$\mathbf{R}_B^* = \mathbf{R}_B - \mathbf{K}_{BI}(1)\mathbf{V}_I(1) - \mathbf{K}_{BI}(2)\mathbf{V}_I(2) - \ldots - \mathbf{K}_{BI}(q)\mathbf{V}_I(q) \ldots$$

As demonstrated above, the static condensation step collapses the effect of the linear equations associated with the non-interface degrees of freedom onto those associated with the interface degrees of freedom. This reduces the number of linear simultaneous equations to be solved to those associated with the interface only. The algorithm for concurrent reduction of the subdomains is presented in Table 12.

TABLE 12
Concurrent reduction of the subdomains

1. Store the interface stiffness matrix $\mathbf{K}_{BB}^*$ with dimensions $n_B \times n_B$ and the interface load vector $\mathbf{R}_B^*$ with dimensions $n_B \times 1$.
2. Distribute the number of internal nodes as evenly as possible among a number of sets equal to the number of prescribed processors. The balance of internal nodes achieved at the end of the final partitioning stage is used to determine these sets.
3. Create a number of threads equal to the number of prescribed processors. Each set is mapped onto a thread.
4. For $q = 1, \ldots,$ *num_threads*, do <u>concurrently</u>
   (a) Factorize

   $$\mathbf{K}_{II}(q) = \mathbf{L}_{II}(q)\mathbf{D}_{II}(q)\mathbf{L}_{II}^T(q)$$

   Assuming the dimensions of the matrix $\mathbf{K}_{II}(q)$ to be $n_q \times n_q$, the elements of $\mathbf{L}_{II}(q)$ and $\mathbf{D}_{II}(q)$ are obtained as follows
   For $j = 1, \ldots, n_q$, do <u>sequentially</u>
     For $i = m_j, \ldots, j-1$, do <u>sequentially</u>

   $$[\mathbf{L}_{II}(q)]_{ij} = \frac{[\mathbf{K}_{II}(q)]_{ij} - \sum_{g=m_m}^{i-1} [\mathbf{L}_{II}(q)]_{gi}[\mathbf{L}_{II}(q)]_{gj}[\mathbf{D}_{II}(q)]_{gg}}{D_{ii}}$$

   Next i

   $$[\mathbf{D}_{II}(q)]_{jj} = [\mathbf{K}_{II}(q)]_{jj} - \sum_{g=mj}^{j-1} [\mathbf{L}_{II}(q)]_{gj}[\mathbf{L}_{II}(q)]_{gj}[\mathbf{D}_{II}(q)]_{gg}$$

   Next j
   m_i is the row number of the first nonzero element in column i,
   m_j is the row number of the first nonzero element in column j, and
   m_m is the maximum of m_i and m_j.
   (b) Solve

   $$[\mathbf{L}_{II}(q)\mathbf{D}_{II}(q)\mathbf{L}_{II}^T(q)]\mathbf{U}_{IB}(q) = \mathbf{K}_{IB}(q) \text{ for } \mathbf{U}_{IB}(q)$$

   The following equations are used for forward and backward substitutions, respectively.
   For $i = 1, \ldots, n_q$, do <u>sequentially</u>
     For $j = 1, \ldots, n_B$, do <u>sequentially</u>

   $$[\mathbf{D}_{II}(q)\mathbf{L}_{II}^T(q)\mathbf{U}_{IB}(q)]_{ij}$$
   $$= [\mathbf{K}_{IB}(q)]_{ij} - \sum_{g=1}^{i-1} [\mathbf{L}_{II}(q)]_{gi}$$
   $$\times [\mathbf{D}_{II}(q)\mathbf{L}_{II}^T(q)\mathbf{U}_{IB}(q)]_{gj}$$

   Next j

Concurrent Analysis of Structures

TABLE 12—contd.

Next i
For $i = n_q, \ldots, 1$, do sequentially
For $j, \ldots, n_B$, do sequentially

$[\mathbf{U}_{IB}(q)]_{ij}$

$$= \frac{[\mathbf{D}_{II}(q)\mathbf{L}_{II}^T(q)\mathbf{U}_{IB}(q)]_{ij} - \sum_{g=i+1}^{n_q} [\mathbf{D}_{II}(q)]_{ii}[\mathbf{L}_{II}(q)]_{ig}[\mathbf{U}_{IB}(q)]_{gj}}{[\mathbf{D}_{II}(q)]_{ii}}$$

Next j
Next i

(c) Solve

$$[\mathbf{L}_{II}(q)\mathbf{D}_{II}(q)\mathbf{L}_{II}^T(q)]\mathbf{V}_I(q) = \mathbf{R}_I(q) \quad \text{for} \quad \mathbf{V}_I(q)$$

The following equations are used for forward and backward substitutions, respectively.
For $i = 1, \ldots, n_q$, do sequentially

$[\mathbf{D}_{II}(q)\mathbf{L}_{II}^T(q)\mathbf{V}_I(q)]_i$

$$= [\mathbf{R}_I(q)]_i - \sum_{g=1}^{i-1} [\mathbf{L}_{II}(q)]_{gi}[\mathbf{D}_{II}(q)\mathbf{L}_{II}^T(q)\mathbf{V}_I(q)]_g$$

Next i
For $i = n_q, \ldots, 1$, do sequentially

$[\mathbf{V}_I(q)]_i$

$$= \frac{[\mathbf{D}_{II}(q)\mathbf{L}_{II}^T(q)\mathbf{V}_I(q)]_i - \sum_{g=i+1}^{n_q} [\mathbf{D}_{II}(q)]_{ii}[\mathbf{L}_{II}(q)]_{ig}[\mathbf{V}_I(q)]_g}{[\mathbf{D}_{II}(q)]_{ii}}$$

Next i

(d) Evaluate

$$\mathbf{K}_{BB}^*(q) = -\mathbf{K}_{IB}^T(q)\mathbf{U}_{IB}(q)$$

(e) Evaluate

$$\mathbf{R}_B^*(q) = -\mathbf{K}_{IB}^T(q)\mathbf{V}_I(q)$$

(f) Update the interface stiffness matrix $\mathbf{K}_{BB}^*$ and interface load vector $\mathbf{R}_B^*$ through the following two equations, respectively (synchronization is required):

$$\mathbf{K}_{BB}^* = \mathbf{K}_{BB}^* + \mathbf{K}_{BB}^*(q)$$
$$\mathbf{R}_B^* = \mathbf{R}_B^* + \mathbf{R}_B^*(q)$$

Note that a racing condition is encountered while updating the interface stiffness matrix and load vector in step 4(f) of the algorithm outlined in Table 12. The two strategies mentioned earlier in assembling the structure stiffness matrix are employed to overcome such a racing condition. In the first strategy, each location of the interface stiffness matrix (only the upper triangle is stored owing to symmetry) and the interface load vector is safeguarded with a semaphore of its own to ensure that access to this location by the processors is done mutually exclusive in time. In the second strategy, extra storage locations are used to store the components of the locations susceptible to a racing condition (all locations of the interface stiffness matrix and load vector). After the static condensation step has been completed, a new set of threads is created to add up the components of the overlapped locations and store the results in the appropriate locations of the interface stiffness matrix and load vector, for this second strategy. A comparison between the two strategies is given in Section 4.10.

4.7 INTERFACE LINEAR EQUATIONS

In this section, concurrent solution of the set of interface linear equations defined by eqn (22) is presented. The same three basic stages mentioned in Section 4.6 are used here, with the exception that the first stage of factorization is replaced by an explicit pivoting procedure in order to transform the interface stiffness matrix to an upper triangular matrix. A better workload balance is achieved with the explicit pivoting approach than the factorization approach.

Table 13 summarizes the algorithm for this step. The number of tasks required in each of the three main stages involved, namely, transforming the matrix to an upper triangle, and forward and backward substitutions is 3 times the *number of interface nodes*. In each task of the first stage, all the terms below the diagonal term (pivoting term) in a given column of the interface stiffness matrix are made zero. Figure 55(a) shows the workload balance attained among the processors when the term (45, 45), for example, is the pivoting element, for the case of three processors (and three threads). The number in each square represents the number of the thread assigned to update this location. Similarly, Figures 55(b) and (c) show the workload balance attained in the calculations involved when finding

TABLE 13
Concurrent solution of interface linear equations

Define the interface stiffness matrix $\mathbf{K}_{BB}^*$ with dimensions $n_B \times n_B$, the interface load vector $\mathbf{R}_B^*$ with dimensions $n_B \times 1$, and the interface displacements vector $\mathbf{r}_B^*$ with dimensions $n_B \times 1$.

Upper triangularization
For $i = 1, \ldots, n_B$, do sequentially
1. Determine the pivot element $[\mathbf{K}_{BB}^*]_{ii}$.
2. For $j = 1, \ldots, num\_threads$, do concurrently
   DO
   (a) Determine the location of the element of the interface stiffness matrix to be updated, for example, the element in the pth row and qth column, $[\mathbf{K}_{BB}^*]_{pq}$ ($p \leq q$).
   (b) Update this element using the relation

$$[\mathbf{K}_{BB}^*]_{pq} = [\mathbf{K}_{BB}^*]_{pq} - \frac{[\mathbf{K}_{BB}^*]_{ip}}{[\mathbf{K}_{BB}^*]_{ii}}[\mathbf{K}_{BB}^*]_{iq}$$

   (c) Update the location: location = location + $num\_threads$ (see Fig. 55a).
   WHILE (location is within the bounds of the skyline array)
Next i

Forward substitution
For $i = 1, \ldots, n_B$, do sequentially
1. Determine the ith element of the updated interface load vector, $[\mathbf{R}_B^*]_i$
2. For $j = 1$ until $num\_threads$, do concurrently
   DO
   (a) Determine the location of the element of the interface load vector to be updated, for example, the element in the pth row, $[\mathbf{R}_B^*]_p$ ($i < p$).
   (b) Update this element using the relation

$$[\mathbf{R}_B^*]_p = [\mathbf{R}_B^*]_p - \frac{[\mathbf{K}_{BB}^*]_{ip}}{[\mathbf{K}_{BB}^*]_{ii}}[\mathbf{R}_B^*]_i$$

   (c) Update the location: location = location + $num\_threads$ (see Fig. 55b).
   WHILE (location is within the bounds of the interface array)
Next i

Backward substitution
For $i = n_B, \ldots, 1$, do sequentially
1. Evaluate the ith element of the interface displacement vector, $[\mathbf{r}_B^*]_i$:

$$[\mathbf{r}_B^*]_i = \frac{[\mathbf{r}_B^*]_i}{[\mathbf{K}_{BB}^*]_{ii}}$$

(*continued*)

TABLE 13—contd.

2. For $j = 1, \ldots, num\_threads$, do <u>concurrently</u>
   DO
   (a) Determine the location of the element of the interface displacement vector to be updated, for example, the element in the qth row, $[\mathbf{r}_B^*]_q$ ($i > q$).
   (b) Update this element using the relation

   $$[\mathbf{r}_B^*]_q = [\mathbf{r}_B^*]_q - [\mathbf{K}_{BB}^*]_{qi}[\mathbf{r}_B^*]_i$$

   (c) Update the location: location = location + $num\_threads$ (see Fig. 55c).
   WHILE (location is within the bounds of the interface array)
   Next i

the 45th term of the adjusted load vector and the 45th term of the displacement vector for the forward and backward substitutions stages, respectively, for the case of three processors.

The only hindrance to the workload balance attained for each of the three stages is that some threads will become idle one after another at the steps nearing the end of each stage. This will have some influence on the efficiency of the concurrent processing. Note, however, the number of times a set of threads is created in this solution approach. The total number of sets of threads required for solving the interface linear equations is equal to 9 times the *number of interface nodes* (each set has a number of threads equal to the number of prescribed processors). Compare this number with the one set of threads required for accomplishing the entire static condensation step using the first strategy or the total of two sets when the second strategy is in effect. This shows the merit of the overall strategy used in this work in reducing the number of linear equations to be solved simultaneously from 3 times the *number of nodes* to 3 times the *number of interface nodes*, thus minimizing the inherently sequential or 'fine-grained parallelism' portion of the solution process. Also, note that the Gaussian elimination procedure has an advantage over other techniques for solving linear equations such as the Givens algorithm (Bathe [1982]), where the number of sets of threads required to transform the interface stiffness matrix to an upper triangular matrix is of the order of (*number of interface nodes*)$^2/2$ and the entire interface stiffness matrix needs to be stored (Adeli & Kamal [1988]).

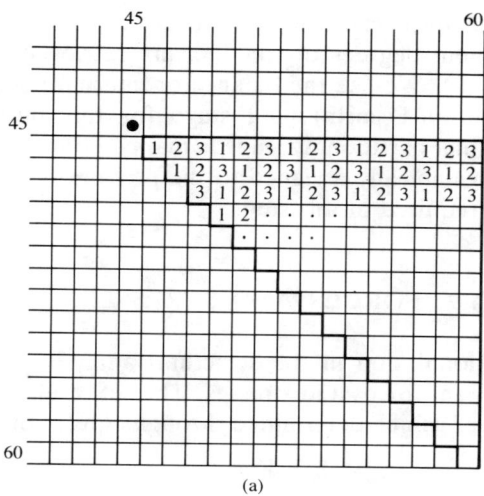

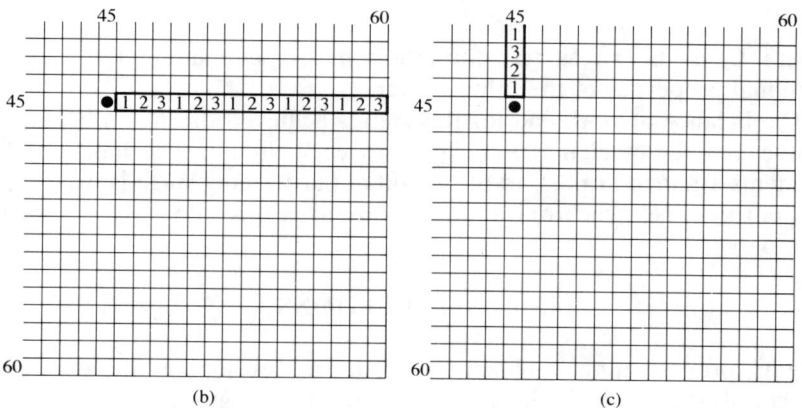

Fig. 55. Illustration of parallelism in the stages involved in the simultaneous solution of the interface linear equations (case of 3 processors): (a) upper triangularization; (b) forward substitution; (c) backward substitution.

4.8 NON-INTERFACE DISPLACEMENTS

Once the interface displacements have been determined, the non-interface displacement degrees of freedom are retrieved through the use of eqns (18)–(20). We use the balance of internal nodes achieved at the end of the final partitioning stage for this purpose. Each subdomain of the final partitioning stage is mapped onto a thread. The threads are then dynamically assigned to the processors. Note that no synchronization is required in this step.

4.9 FORCES AND STRESSES

The last computational step in the structural analysis problem is to determine the element forces and stresses. The set of element internal forces S_i in member i can be determined through eqn (2) of Section 2.4:

$$S_i = K_i r_i \qquad (2)$$

where K_i is the i th element stiffness matrix and r_i is the generalized displacement vector for the i th element. Finally, the stress vector σ_i of the i th element can be determined from eqn (3) of Section 2.4:

$$\sigma_i = B_i S_i \qquad (3)$$

where B_i is the stress transformation matrix containing the cross-sectional properties of the i th member.

The balance of workload for this step is achieved through a balance of the number of elements assigned to each thread. The balance of elements attained at the end of the initial partitioning stage is used for this purpose. No synchronization is required in this step.

4.10 APPLICATIONS

The C programming language is used for developing the parallel structural analysis code. The functions of this program are shown in Fig. 56. The program has been applied to several truss and frame structures for a variable number of processors on the Encore Multimax shared-memory multiprocessor computer. A minimum of nine elements per subdomain is prescribed as a pre-condition for the initial partitioning stage. Similarly, a minimum of three internal nodes per

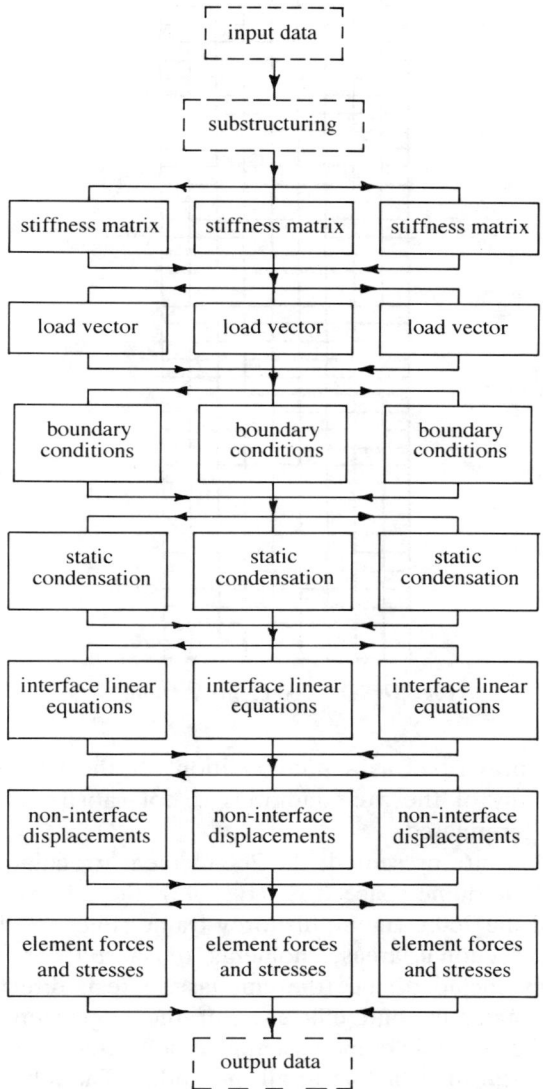

Fig. 56. Flow chart of the parallel C program for analysis of framed structures.

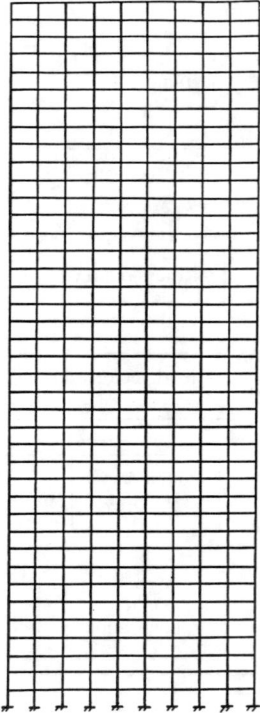

Fig. 57. The 760-element frame.

subdomain is prescribed as a pre-condition for the final partitioning stage. When any of the pre-conditions is not satisfied, execution is automatically terminated.

Four examples are presented: the 266-element irregular frame (Fig. 48), the geodesic dome space truss (Fig. 42), the 200-bar plane truss (Fig. 45), and the 760-element 40-storey frame (Fig. 57). Information such as cross-sectional areas, moments of inertia, or modulus of elasticity is not included since the emphasis here is directed towards assessing the speed-up and efficiency of the concurrent processing algorithms. The geodesic dome structure is subjected to a set of equal downward vertical loads acting at all the nodes. The left-hand side of the 200-bar plane truss is subjected to horizontal loads acting at the nodes from left to right. Similarly, the left-hand side of the 266- and 760-element frames are subjected to horizontal loads acting at the nodes from left to right in addition to uniformly distributed loads acting on all horizontal members of the frames.

The speed-ups attained in the major computational steps, namely, assembly of the structure stiffness matrix, static condensation, solution of the interface linear equations, retrieving the non-interface nodal displacements, and evaluation of the element forces and stresses, are studied for a variable number of processors. The speed-ups for the steps of assembling the load vector and applying the boundary conditions are not included, since the contribution of each of these steps to the total sequential time for the examples studied in this research is always less than 0·3%. Also, the sequential time consumed in the steps dealing with the solution of the overall set of linear equations, namely, static condensation, solution of the interface linear equations, and non-interface displacements, amounts to 90% of the total sequential time or more in most of the cases studied.

We define the speed-up attained in a certain step of the solution process for a specified number of processors as the ratio of the time spent on executing the sequential code of this step to the time spent on executing the concurrent code, when the specified number of processors is in use. Furthermore, the issue of workload balance is studied. By workload balance, we mean the speed-up when the overhead time required for creating the threads is neglected. This is useful in assessing the effectiveness of the workload balancing strategies developed in this chapter. Also, the overall speed-up, workload balance, and efficiency of the algorithms are assessed. Efficiency is defined as the percentage of the speed-up divided by the number of processors used for the case of workload balance.

4.10.1 Assembling the structure stiffness matrix

The element stiffness matrices are calculated and assembled into the structure stiffness matrix in this step. The balance of elements attained at the end of the initial partitioning stage of the substructuring algorithm is used in this step. As described earlier, two strategies are used and studied for overcoming the racing conditions that arise during the assembly process. In the first, semaphores are used. In the second, additional threads are created for adding-up the overlapped locations of the stiffness matrix. Figure 58 shows the speed-ups attained for the 266-element frame for three cases: threads and

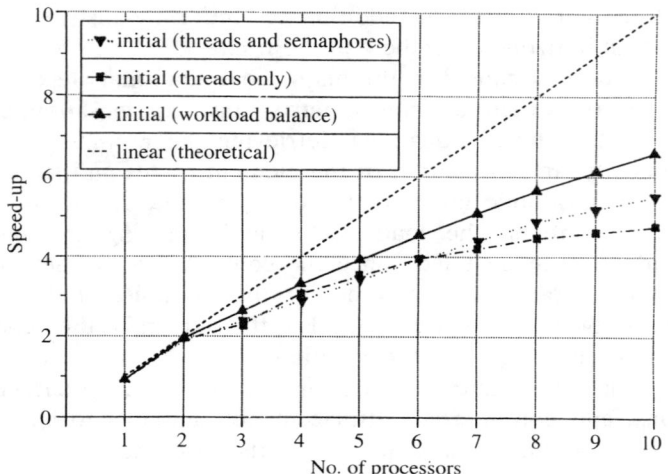

Fig. 58. Speed-ups for assembling the stiffness matrix for the 266-element frame.

semaphores, threads only, and the workload balance (i.e. when the overhead time for creation of threads is neglected).

The 'threads and semaphores' curve represents the case when threads are created for setting up the element stiffness matrices and assembling them into the structure stiffness matrix. In this case, the assembly process is intervened with synchronization wherever necessary. Rather than synchronization, extra storage locations are provided for the overlapped locations in the 'threads only' case. A new set of threads is then created to determine the values of the overlapped locations in the structure stiffness matrix through an extra step of addition. The 'workload balance' case is similar to the 'threads only' case with the exception that the overhead time for creating the threads is neglected. The offset of this curve from the theoretical linear speed-up case is mostly due to presence of the extra addition step, which does not exist in the sequential code.

Because of the small number of overlapped locations in this step (6 times the *number of interface nodes*), the two curves 'threads and semaphores' and 'threads only' representing the two strategies used to overcome the racing condition are not substantially different. For a fixed number of processors, however, the greater the number of overlapped locations, the worse is the speed-up for the case of semaphores.

4.10.2 Static condensation

In this step, the effect of the linear equations associated with the non-interface degrees of freedom is collapsed onto those associated with the interface. Three basic stages are involved: factorizing the subdomain stiffness matrices, reducing the subdomain stiffness matrices coupling the subdomains with the interface and the subdomain load vectors, and updating the interface stiffness matrix and the interface load vector. The percentages of the time spent by the sequential code on executing each of the three stages to the total sequential time required for the static condensation step for the 266-element frame are shown in Fig. 59 for a variable number of processors. The dominance of the 'reduction' stage is apparent for cases with more than one processor. As the number of processors

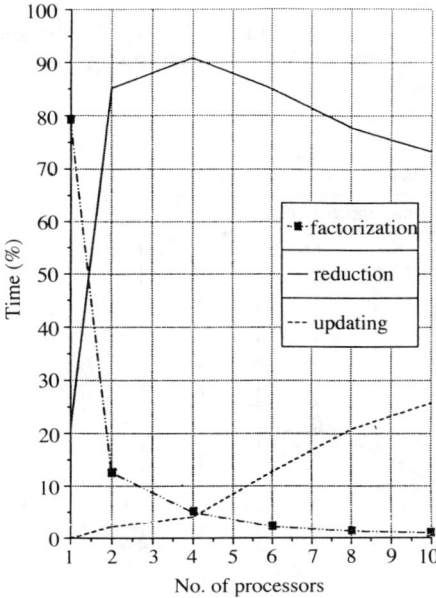

Fig. 59. Time analysis for the static condensation step for the 266-element frame.

98   Parallel Processing in Structural Engineering

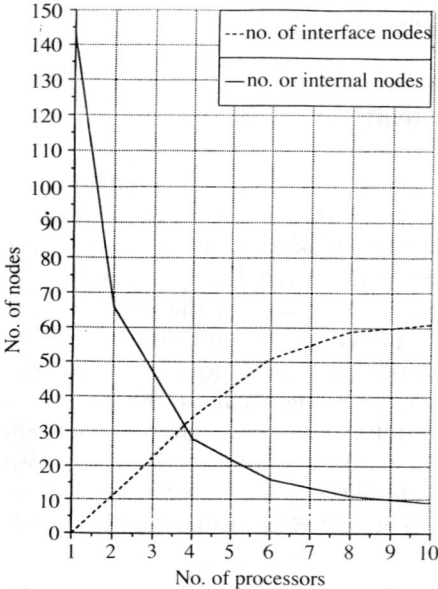

Fig. 60. Numbers of interface and internal nodes for the 266-element frame at the end of the final partitioning stage.

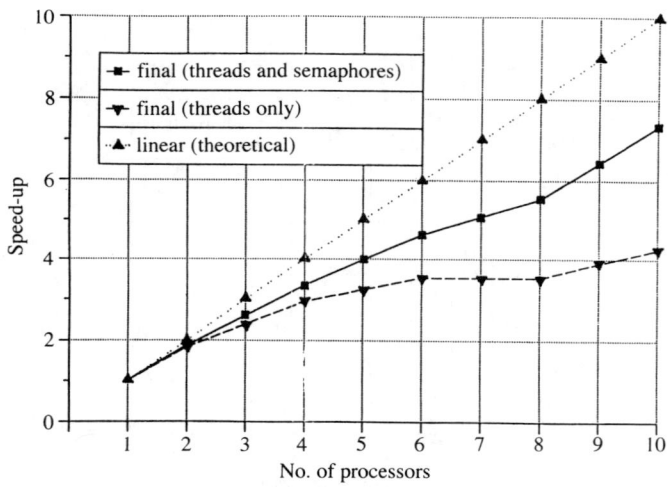

Fig. 61. Speed-ups for the static condensation step for the 266-element frame using different strategies for avoiding a racing condition.

increase, the time consumed in 'updating' increases. This is due to the fact that the number of interface nodes increases as the number of subdomains (and, consequently, processors) increases. Figure 60 shows the number of internal nodes assigned to each subdomain and the number of interface nodes at the end of the final partitioning stage for the 266-element frame for various numbers of processors in use.

The two aforementioned strategies for avoiding a racing condition in the 'updating' stage are compared. Figure 61 compares the speed-ups attained for the cases of 'threads and semaphores' and 'threads only', using the final partitioning stage for the 266-element frame. Better speed-ups are achieved for the case of 'threads only' compared with the case of 'threads and semaphores'. The difference becomes larger as the number of processors increases. This is due to the fact that the number of interface nodes increases as the number of processors increases (Fig. 60). Hence, the number of sections that need to be safeguarded against a racing condition during updating the interface arrays increases, and more semaphores are required. The overhead time required for the invocation of such semaphores is greater than that required for introducing an extra addition step and creating a set of threads to parallelize this step in the 'threads only' case.

The overall time required for creating the threads is insignificant in

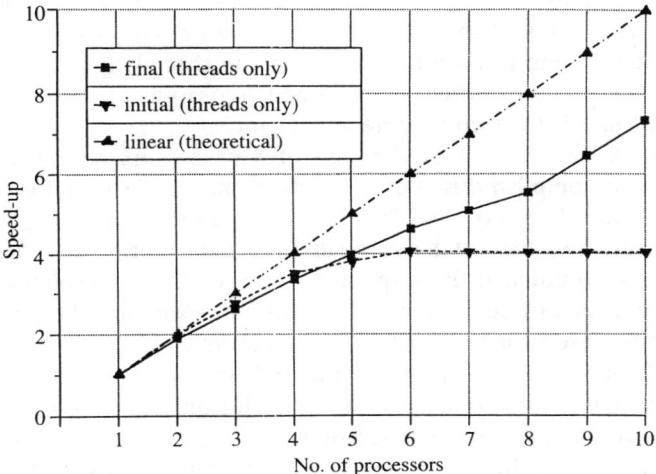

Fig. 62. Speed-ups for the static condensation step for the 266-element frame at the end of the initial and final partitioning stages.

this step compared with the overall computational time required for the static condensation step. Therefore, the workload balance curve practically coincides with the 'threads only' curve. Most of the offset of the workload balance curve from the theoretical linear speed-up curve is due to the differences in bandwidths of the subdomains matrices. This affects the speed-ups mainly in the 'reduction' stage of the static condensation step. The rest of the offset is due to the introduction of the extra addition step.

In order to assess the efficiency of the substructuring algorithm developed in Chapter III, Fig. 62 shows a comparison of the balance of elements achieved at the end of the initial partitioning stage with the balance of internal nodes achieved at the end of the final partitioning stage for the 266-element frame. This figure demonstrates that a substantial improvement in the speed-up is obtained at the end of the final partitioning stage.

4.10.3 Solution of interface linear equations

This step solves the linear equations associated with the interface degrees of freedom. This step consists of three stages: transforming the interface stiffness matrix to an upper-triangular one, and forward and backward substitutions. Each stage requires a number of sets of threads equal to 3 times the *number of interface nodes,* with a total number of sets equal to 9 times the *number of interface nodes.* Since the overhead time required for creating the threads is added sequentially to the execution time of the concurrent application, its effect on the speed-up of this step is quite significant.

As an example, consider the case of 5 processors and 30 interface nodes. To accomplish this step, 270 sets of threads, each set containing 5 threads, are then required. If the time required for creating a thread is one unit then the total overhead time for creating the threads is $270 \times 5 = 1350$ units. If the sequential time required for solving the set of interface linear equations is 1200 units, for example, the concurrent time for this step for the case of 5 processors is $1200/5 + 1350 = 1590$ units, which is greater than the sequential time.

At this point, we can draw several conclusions. The step of solution of the interface linear equations suffers from 'fine-grained parallelism', thus necessitating the creation of a large number of threads. The overhead time required for this sets back the speed-up to a great extent. Better overall concurrent performance is obtained when the sequential time required for solving the interface linear equations is a small percentage of the total sequential time and/or the overhead time required for creating the threads for this step is small compared with the total sequential time required for solving the interface linear equations. To relieve some of the burden encountered in this bottleneck situation, however, it is suggested that fewer processors be used than for the rest of the problem. In other words, the program requests a fewer number of processors for executing this step. Considering the previous example, for instance for the case of only two processors (or threads), the total concurrent time is $1200/2 + 540 = 1140$ units, compared with 1590 units when five processors are in use. Heuristic rules can be developed for determining the appropriate number of processors for this step. In this work, the step of solving the interface linear equations is performed several times, each time with a different number of processors less than or equal to the number of processors used for the rest of the problem, and the number of processors corresponding to the minimum concurrent time is chosen. Figure 63 shows the speed-ups attained for the 266-element frame in this step.

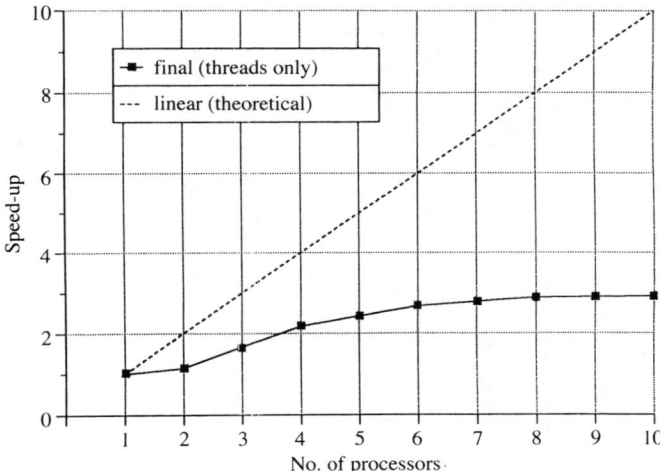

Fig. 63. Speed-up for the solution of interface linear equations for the 266-element frame.

4.10.4 Computation of non-interface nodal displacements

Once the displacements of the interface degrees of freedom have been obtained, the displacements corresponding to non-interface degrees of freedom are found. No synchronization is required in this step. The balance of internal nodes at the end of the final partitioning stage of the substructuring is used to obtain the required workload balance. In order to assess the efficiency of the substructuring algorithm, however, comparison with the balance of elements attained at the end of the initial partitioning stage is also presented. Figure 64 shows the speed-ups attained for the 266-element frame in this step. The overhead time required for creating the threads is insignificant in this step compared with the overall computational time. Therefore, the workload balance curve is practically identical to the 'threads only' curve. Figure 64 also shows a substantial improvement in the speed-up at the end of the final partitioning stage compared with the initial partitioning, and thus the effectiveness of the substructuring algorithm presented in Chapter III.

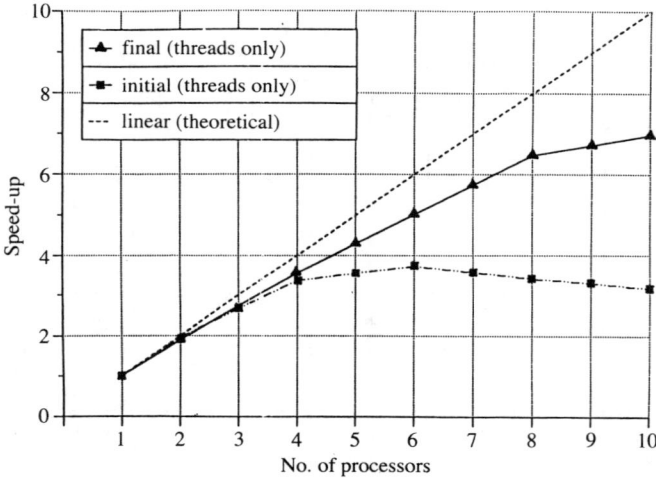

Fig. 64. Speed-ups for computation of the non-interface displacements for the 266-element frame.

4.10.5 Computation of element forces and stresses

The final step is to evaluate the element forces and stresses. The load balance at the end of the initial partitioning stage of the substructuring algorithm is used in this step. No synchronization is required. The cases of 'threads only' and workload balance for the 266-element frame are shown in Fig. 65. The speed-up for the case of workload balance is practically equal to the number of processors used, confirming that computations of the forces and stresses are identical for all the elements.

4.10.6 Overall speed-up, workload balance, and efficiency

The overall performance of the algorithms is assessed in this section. The overall speed-up of all seven steps (namely, assembling the structure stiffness matrix, assembling the load vector, applying the boundary conditions, performing the static condensation, solution of the interface linear equations, finding the non-interface nodal displacements, and evaluating the element forces and stresses) for the case of 'threads only' for the 266-element frame is shown in Fig. 66. The 'initial' and 'final' curves refer to the cases when the steps of static

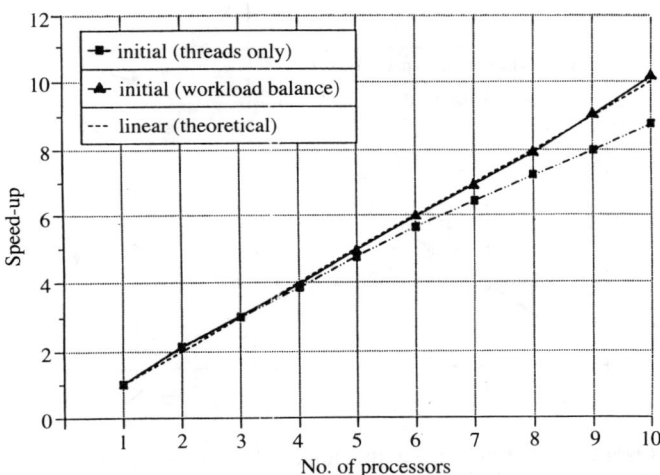

Fig. 65. Speed-up for evaluation of the forces and stresses for the 266-element frame.

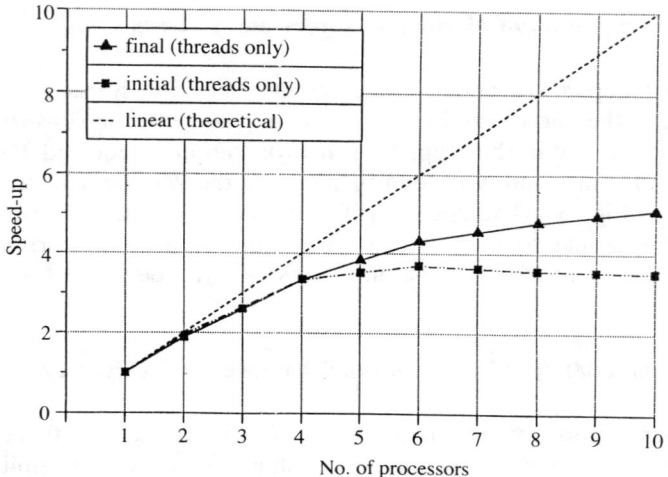

Fig. 66. Overall speed-ups for the 266-element frame.

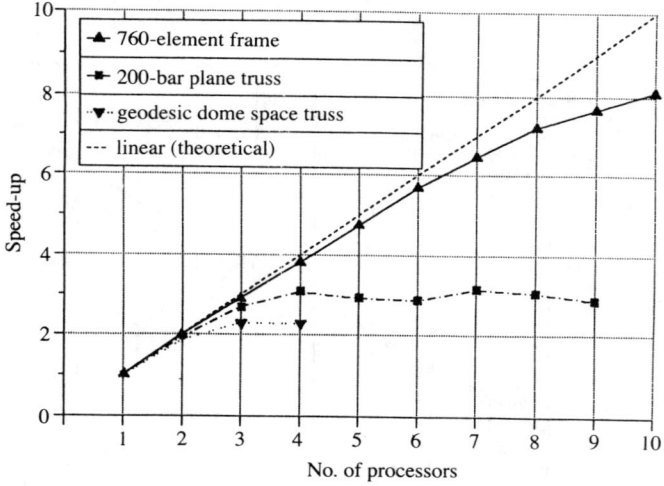

Fig. 67. Overall speed-ups using the final subdomains for the case 'threads only' (for the geodesic dome space truss, 200-bar plane truss, and 760-element frame).

condensation, solution of interface linear equations, and computation of non-interface nodal displacements are performed using the balance of elements achieved at the end of the initial partitioning stage and the balance of internal nodes achieved at the end of the final partitioning stage, respectively. The workload balance for the remaining steps is achieved as described in the previous sections. Better speed-ups are achieved for the case of final partitioning. The effect of the step of solution of the interface linear equations on reducing the speed-up is evident.

Figure 67 shows the speed-up curves for the final partitioning stage for the geodesic dome space structure, the 200-bar plane truss, and the 760-element plane frame. Execution is automatically terminated when the minimum number of elements (nine) or internal nodes (three) is not satisfied within the subdomains. This represents the case of five or more subdomains for the geodesic dome space structure and ten or more subdomains for the 200-bar plane truss. Better speed-ups are achieved when the percentage of the sequential time required for the step of solving the interface linear equations is small compared with the total sequential time and/or the overhead time required for creation of threads in this step is small compared with the time required for accomplishing this step.

Tables 14, 15, and 16 summarize the speed-up for the four examples considered for the cases of 4, 9, and 10 processors, respectively, for the case of final partitioning. The degrees of freedom for each structure are also included in these tables. These results clearly demonstrate that better speed-ups are achieved for larger structures. On one hand, this is due to the fact that the overhead time required for creating a thread tends to be insignificant compared with the time

TABLE 14
Overall performance for the case of four processors

Example	Number of degrees of freedom	Speed-up	Workload balance	Efficiency
Geodesic dome space truss	111	2·27	3·29	82·3
200-bar plane truss	150	3·07	3·62	90·5
266-element frame	429	3·32	3·70	92·5
760-element frame	1200	3·81	3·82	95·5

TABLE 15
Overall performance for the case of nine processors

Example	Number of degrees of freedom	Speed-up	Workload balance	Efficiency
Geodesic dome space truss	111	—	—	—
200-bar plane truss	150	2·87	6·43	71·4
266-element frame	429	4·95	7·04	78·2
760-element frame	1200	7·67	8·26	91·8

required for the thread to execute its task as the size of the task assigned to each thread increases. On the other hand, the percentage of the time consumed in the fine-grained parallelism step (solution of the linear interface equations) tends to be less significant compared with the overall execution time as the size of the problem increases. This explains why in Fig. 67 the speed-up curve for the 760-element frame is substantially higher than the speed-up curve for the 200-bar plane truss, which in turn is higher than that of the geodesic dome space truss. When the overhead time required for creating the threads and the time required for solving the set of linear interface equations become more significant compared with the execution time, the concurrent application fails to pick up additional speed-up as the number of processors increases. This is observed in Fig. 67 for the geodesic dome space truss for the case of four processors and for the 200-bar plane truss for the cases of more than four processors.

The overall workload balance (ignoring the time required for thread creation in the 'threads only' case) for the 266-element frame for the

TABLE 16
Overall performance for the case of ten processors

Example	Number of degrees of freedom	Speed-up	Workload balance	Efficiency
Geodesic dome space truss	111	—	—	—
200-bar plane truss	150	—	—	—
266-element frame	429	5·11	7·80	78·0
760-element frame	1200	8·12	8·95	89·5

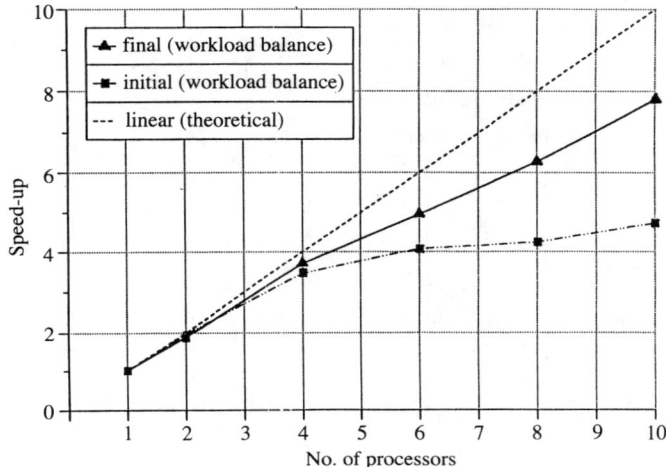

Fig. 68. Overall workload balance for the 266-element frame.

cases of initial and final partitioning is shown in Fig. 68. Figure 69 shows the overall workload balance for the final partitioning stage for the geodesic dome space structure, the 200-bar plane truss, and the 760-element plane frame. The offset of the curves in Figs 68 and 69 from the theoretical linear speed-up case is mainly due to the imbalance in workload. The effect of this imbalance becomes more significant as the number of threads (and processors) increases, and, consequently, the task assigned to each thread becomes smaller.

Similarly, the overall efficiency for the case of initial and final partitioning for the 266-element frame is shown in Fig. 70. Figure 71 shows the overall efficiency for the final partitioning stage for the geodesic dome space structure, the 200-bar plane truss, and the 760-element plane frame. Tables 14, 15, and 16 also include the workload balance and efficiency for the four examples considered in this chapter for the case of final partitioning, using 4, 9, and 10 processors, respectively.

4.11 SUMMARY AND CONCLUSIONS

In this chapter, we have presented parallel algorithms for the complete analysis of framed structures on shared-memory multiprocessor computers such as the Encore Multimax. The algorithms have been

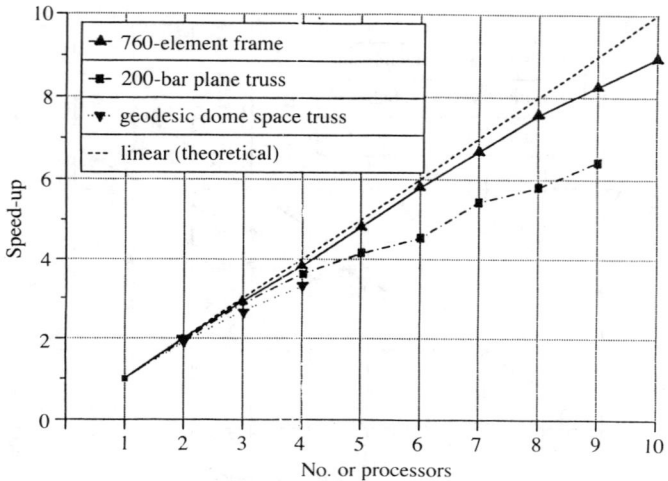

Fig. 69. Overall workload balance for the final subdomains (for the geodesic dome space truss, 200-bar plane truss, and 760-element frame).

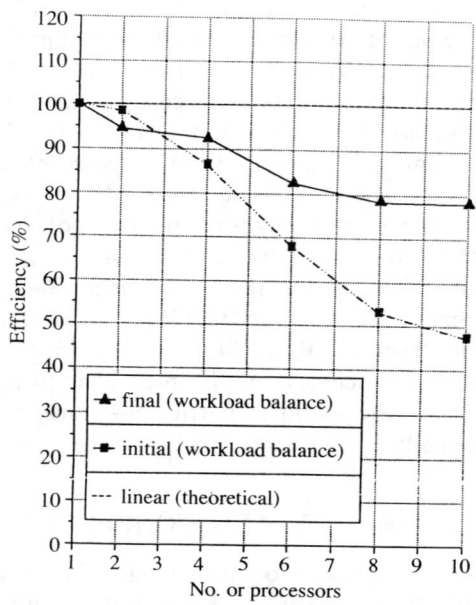

Fig. 70. Overall efficiency for the 266-element frame.

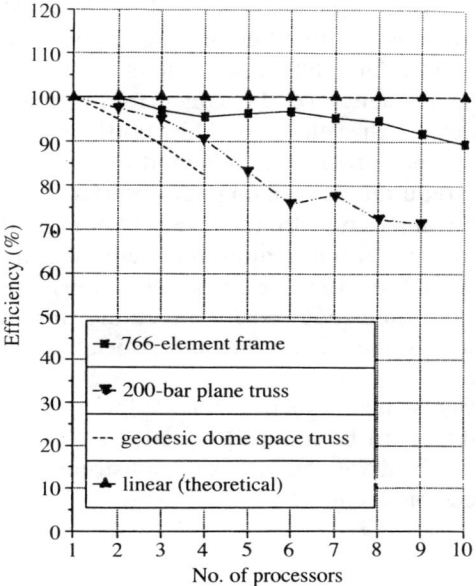

Fig. 71. Overall efficiency for the final subdomains (for the geodesic dome space truss, 200-bar plane truss, and 760-element frame).

implemented in the programming language C on an Encore Multimax shared-memory multiprocessor computer, and have been applied to several truss and frame structures. Emphasis is directed towards attaining a workload balance during each computational step of the solution process. A three-stage substructuring algorithm is used to balance the workload through some of the steps of the solution process. Balancing the internal nodes proves to be efficient for the steps of static condensation and calculation of non-interface displacements. However, balancing the elements is beneficial for the steps of setting up the structure stiffness matrix and calculation of the element forces and stresses. The approach of switching between two different partitioning techniques—'initial partitioning' based on a balance of elements and 'final partitioning' based on a balance of internal nodes within the subdomains—results in an efficient concurrent paradigm.

Two different strategies have been studied for avoiding a racing condition: synchronization of threads through the use of semaphores and introduction of an extra addition step and creation of a set of threads to execute this step concurrently. The second strategy becomes more advantageous as the number of overlapped locations increases. Solution of the interface linear equations represents a bottleneck situation for the efficiency of the concurrent application. For large structures, however, the effect of this bottleneck situation is less severe, and better performance is obtained because the percent of the sequential time required for solving the interface linear equations is generally small compared with the overall sequential time, and the percentage of the overall time required for creating the threads for the step of solution of the interface linear equations is small compared with the sequential time required for executing this step. An alternative that alleviates some of the burden associated with this step has been presented.

Efficient storage schemes such as the skyline and the compact form allow large problems to be handled effectively. Overall efficiencies of 90–100% are achieved for the case of a 40-storey 760-element frame with 1200 degrees of freedom.

Chapter V

Concurrent Optimization of Structures

5.1 INTRODUCTION

In this chapter, we present parallel algorithms for optimization of framed structures subjected to multiple loading conditions and displacement and stress constraints. This is done in the context of the optimality criteria approach. Since the solution of the structural optimization problem is essentially iterative, a considerable amount of computational efficiency can be achieved if parallelism is introduced at the iteration level. Each iteration is composed of two phases: analysis and redesign, as shown in Fig. 72.

The partitioning of structures described in Chapter III is succeeded by seven steps using the stiffness (displacement) method of structural analysis, as detailed in Chapter IV. These steps are as follows: assembling the structure stiffness matrix, setting up the load vector, applying the boundary conditions, condensation of the non-interface displacement degrees of freedom onto the interface ones, solution of the interface displacement degrees of freedom, evaluation of the non-interface displacements, and computation of element forces and stresses (Fig. 72). Continuing on to the redesign phase, we find that some steps of the analysis phase are re-visited, while other new steps are introduced (Fig. 72).

The rest of the chapter is organized as follows. The equations involved in the optimality criteria approach for optimal structural design are discussed briefly in Section 5.2. The algorithms for parallel structural optimization on shared memory multiprocessor computers are outlined in Section 5.3. Applications are presented for several

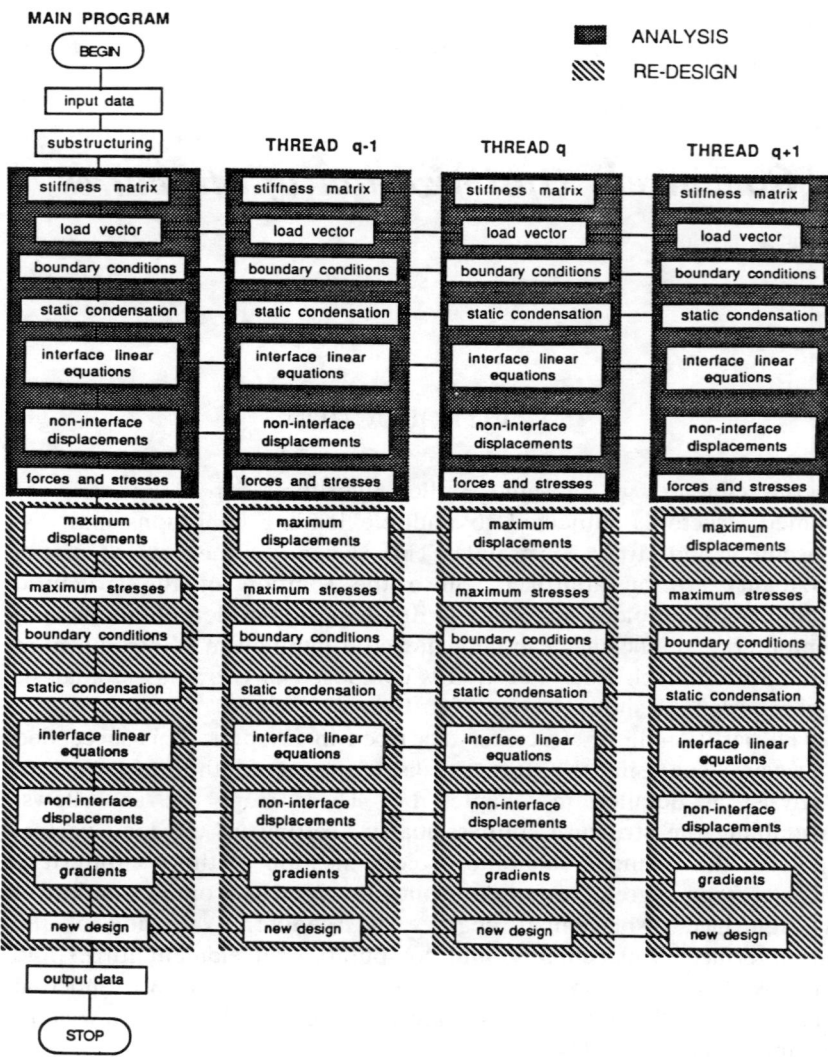

Fig. 72. Macro flow chart of the parallel C program for optimization of framed structures.

truss and frame problems in Section 5.4. Finally, Section 5.5 gives a summary and conclusions.

5.2 AN OPTIMALITY CRITERION APPROACH

In this section, we present an optimality criterion approach for the optimal design of structures subjected to multiple loading conditions. Basically, we combine three different recurrence formulas to come up with an optimization algorithm that is efficient and suitable for parallel processing on shared memory multiprocessor computers. The optimum structural design problem can be cast in the following form: find the vector of design variables, $\mathbf{x}$, such that

$$W(\mathbf{x}) = \sum_{m=1}^{M} \rho_m l_m x_m \to \min \qquad (23)$$

subject to

$$r_p^l \leq r_{pg}(\mathbf{x}) \leq r_p^u \quad (p=1,2,\ldots,P; g=1,2,\ldots,G) \qquad (24)$$

$$\sigma_m^l \leq \sigma_{mg}(\mathbf{x}) \leq \sigma_m^u \quad (m=1,2,\ldots,M; g=1,2,\ldots,G) \qquad (25)$$

$$x_v^l \leq x_v \leq x_v^u \quad (v=1,2,\ldots,V) \qquad (26)$$

where $W(\mathbf{x})$ is the objective function represented by the weight of the structure. The terms ρ_m, l_m, and x_m represent the specific weight, the length, and the cross-sectional area of the mth element, respectively. The terms r_{pg}, σ_{mg}, and x_v represent the nodal displacements, member stresses, and design variables, respectively. The constants r_p^l, r_p^u, σ_m^l, σ_m^u, x_v^l, and x_v^u are the lower and upper bounds on the nodal displacements, member stresses, and design variables, respectively. The constants M, P, G, and V represent the number of elements in the structure, the number of constrained displacements, the number of loading cases, and the number of design variables, respectively. Note that the terms 'element' and 'member' are used interchangeably. The two behavior constraints represented by eqns (24) and (25) are considered separately, with the objective function represented by eqn (23), in order to find the recurrence formula for each case. The algorithm that combines the two recurrence formulae with eqns (23) and (26) is presented in Section 5.3.

Considering the displacement constraints alone and assuming that $r_p^u = -r_p^l = \bar{r}_p$, eqn (24) can be written in the form

$$r_{pg}(\mathbf{x}) \leq \bar{r}_p \quad (p=1,2,\ldots,P; g=1,2,\ldots,G) \qquad (27)$$

The Lagrangian function for the optimization problem consisting of eqns (23) and (27) is expressed as

$$L(\mathbf{x}, \lambda) = W(\mathbf{x}) + \sum_{p=1}^{P} \sum_{g=1}^{G} \lambda_{pg}[r_{pg}(\mathbf{x}) - \bar{r}_p] \qquad (28)$$

where λ_{pg} is the Lagrange multipler associated with the pth displacement constraint and the gth loading case.

At this point, we assume that the number of design variables, V, is equal to the number of members in the structure, M. This assumption will be relaxed subsequently. The Kuhn–Tucker necessary conditions for the local constrained optimum are given by

$$\frac{\partial L(\mathbf{x}, \lambda)}{\partial x_m} = \frac{\partial W(\mathbf{x})}{\partial x_m} + \sum_{p=1}^{P} \sum_{g=1}^{G} \lambda_{pg}$$

$$\times \frac{\partial r_{pg}(\mathbf{x})}{\partial x_m} = 0 \quad (m = 1, 2, \ldots, M) \qquad (29)$$

subject to

$$\lambda_{pg}[r_{pg}(\mathbf{x}) - \bar{r}_p] = 0 \quad (p = 1, 2, \ldots, P; g = 1, 2, \ldots, G) \qquad (30)$$

$$\lambda_{pg} \geq 0 \quad (p = 1, 2, \ldots, P; g = 1, 2, \ldots, G) \qquad (31)$$

Examining eqns (30) and (31), we can see that $\lambda_{pg} > 0$ for active constraints ($r_{pg}(\mathbf{x}) = \bar{r}_p$) and that $\lambda_{pg} = 0$ for inactive (loose) constraints ($r_{pg}(\mathbf{x}) < \bar{r}_p$).

It can be shown that (Khot et al. [1979])

$$\frac{\partial r_{pg}(\mathbf{x})}{\partial x_m} = -\frac{\mathbf{r}_{mg}^T \mathbf{K}_m \mathbf{y}_{pm}}{x_m} \qquad (32)$$

where $\mathbf{r}_{mg}^T$ is the transpose of the displacement vector of the mth member due to the gth loading case, $\mathbf{K}_m$ is the stiffness matrix of the mth member of the structure, and $\mathbf{y}_{pm}$ is the displacement vector of member m due to a unit load applied in the direction of the pth constrained displacement. Substituting eqn (32) into eqn (29), we can write eqns (29)–(31) as

$$\rho_m l_m - \sum_{p=1}^{P} \sum_{g=1}^{G} \lambda_{pg} \frac{\mathbf{r}_{mg}^T \mathbf{K}_m \mathbf{y}_{pm}}{x_m} = 0 \quad (m = 1, 2, \ldots, M) \qquad (33)$$

subject to

$$\lambda_{pg}[r_{pg}(\mathbf{x}) - \bar{r}_p] = 0 \quad (p = 1, 2, \ldots, P; g = 1, 2, \ldots, G) \qquad (30)$$

$$\lambda_{pg} \geq 0 \quad (p = 1, 2, \ldots, P; g = 1, 2, \ldots, G) \qquad (31)$$

Suppose now that the αth displacement in the βth loading case is exactly active (i.e. $r_{\alpha\beta}(\mathbf{x}) = \bar{r}_\alpha$ and $\lambda_{\alpha\beta} > 0$), while all other constraints are loose (i.e. the corresponding λ values are equal to zero). In this case, eqn (33) can be expressed as

$$\rho_m l_m x_m - \lambda_{\alpha\beta} \mathbf{r}_{m\beta}^T \mathbf{K}_m \mathbf{y}_{\alpha m} = 0 \quad (m = 1, 2, \ldots, M) \quad (34)$$

Adding up all M equations in (34) and solving for the value of $\lambda_{\alpha\beta}$, we obtain

$$\lambda_{\alpha\beta} = \frac{\sum_{m=1}^{M} \rho_m l_m x_m}{\sum_{m=1}^{M} \mathbf{r}_{m\beta}^T \mathbf{K}_m \mathbf{y}_{\alpha m}} = \frac{W(\mathbf{x})}{r_{\alpha\beta}} = \frac{W}{\bar{r}_\alpha} \quad (35)$$

Substitution of eqn (35) into eqn (34) and rearranging terms yields

$$1 = \left(\frac{W}{\bar{r}_\alpha}\right)\left(\frac{\mathbf{r}_{m\beta}^T \mathbf{K}_m \mathbf{y}_{\alpha m}}{\rho_m l_m x_m}\right), \quad (m = 1, 2, \ldots, M) \quad (36)$$

which is the optimality criterion. Any design that satisfies eqn (36) is at least a local minimum.

Multiplying both sides of eqn (36) by $(x_m)^\eta$ and taking the ηth root of both sides yields the following recurrence formula:

$$\{x_m\}_{\mu+1} = \{x_m\}_\mu \left\{\left[\left(\frac{W}{\bar{r}_\alpha}\right)\left(\frac{\mathbf{r}_{m\beta}^T \mathbf{K}_m \mathbf{y}_{\alpha m}}{\rho_m l_m x_m}\right)\right]^\zeta\right\}_\mu \quad (m = 1, 2, \ldots, M) \quad (37)$$

where μ and $\mu + 1$ are the iteration numbers, and $\zeta = 1/\eta$ determines the step size. The parameter ζ controls the convergence and stability of the method. Experience shows that values of ζ in the range 0·001–0·2 result in optimum designs without difficulty (Khan et al. [1979]). In this work, we start with $\zeta = 0\cdot 1$. If oscillation occurs in the objective function at some iteration, the value of ζ is halved ($\zeta \to \frac{1}{2}\zeta$) before proceeding to the next iteration. In cases where N_v members have the same design variable x_v, eqn (37) is written in the form

$$\{x_v\}_{\mu+1} = \{x_v\}_\mu \left\{\left[\left(\frac{W}{\bar{r}_\alpha}\right)\left(\frac{\sum_{i=1}^{N_v} \mathbf{r}_{i\beta}^T \mathbf{K}_i \mathbf{y}_{\alpha i}}{\rho_v x_v \sum_{i=1}^{N_v} l_i}\right)\right]^\zeta\right\}_\mu \quad (v = 1, 2, \ldots, V) \quad (38)$$

Now suppose that more than one displacement constraint is active at optimality. In this case, the Lagrange multipliers can no longer be evaluated through the use of eqn (35). Equation (33) results in a set of linear equations in terms of the Lagrange multipliers (Cheng et al. [1981]). Solution of this set of equations can be cumbersome. Instead, we use a recurrence formula for estimating the Lagrange multipliers in each iteration. For an active displacement constraint, eqn (27) can be written as an equality (i.e. $r_{pg}(\mathbf{x}) = \bar{r}_p$). Multiplying both sides of this equality by $(\lambda_{pg})^{1/\psi}$, raising both sides to the ψth power, and arranging terms, we obtain the following recurrence formula for estimating the Lagrange multipliers:

$$\{\lambda_{pg}\}_{\mu+1} = \{\lambda_{pg}\}_\mu \left\{ \left[\left(\frac{r_{pg}(\mathbf{x})}{\bar{r}_p} \right) \right]^\psi \right\}_\mu$$

$$= \{\lambda_{pg}\}_\mu \left\{ \left[\left(\frac{r_{pg}}{\bar{r}_p} \right) \right]^\psi \right\}_\mu \quad (39)$$

where μ and $\mu + 1$ are the iteration numbers, while the parameter ψ determines the step size. The value of ψ should be chosen so that it increases the effect of λ corresponding to a constraint of high importance and decreases the effect of λ corresponding to a constraint of low importance. Our experience in this work indicates that a value of $\psi = 2$ results in the optimum design being obtained without difficulty. In addition, we use eqn (35) for finding an initial estimate of the values λ_{pg}. Considering the case of B active displacement constraints, we replace eqn (38) by the following recursion relation:

$$\{x_v\}_{\mu+1} = \{x_v\}_\mu \left\{ \left[\sum_{b=1}^B \left(\lambda_{pg} \frac{\sum_{i=1}^{N_v} \mathbf{r}_{ig}^T \mathbf{K}_i \mathbf{y}_{pi}}{\rho_v x_v \sum_{i=1}^{N_v} l_i} \right) \right]^\zeta \right\}_\mu$$

$$(v = 1, 2, \ldots, V) \quad (40)$$

Now, we consider the stress constraints alone. The Kuhn–Tucker conditions for the optimal solution of eqns (23) and (25) result in the following well-known stress-ratio recursion formula:

$$\{x_v\}_{\mu+1} = \{x_v\}_\mu \left\{ \left[\max_{\text{all } i \in N_v, \text{ all } g \in G} \left(\frac{\sigma_{ig}}{\sigma_i^l \text{ or } \sigma_i^u} \right) \right] \right\}_\mu \quad (v = 1, 2, \ldots, V) \quad (41)$$

The parallel algorithms that combine the recurrence formulas represented by eqn (40) (or eqn 38) and eqn (41) together with eqns (23) and (26) are presented in Section 5.3.

5.3 ALGORITHMS FOR PARALLEL STRUCTURAL OPTIMIZATION

In this section, we present the algorithms for parallel structural optimization of framed structures using the optimality criterion approach outlined earlier. The solution procedure is iterative. Each iteration is composed of an analysis phase followed by a redesign phase. Figure 72 highlights the major steps within each phase. In each step, a number of threads equal to the number of processors specified by the user is created. The computational task involved within each step is automatically decomposed into a number of subtasks equal to the number of threads. Each subtask is mapped onto a thread. For some of the tasks, a racing condition is inevitable and synchronization of the threads is required. Whenever such a situation is encountered, extra storage locations are used to avoid simultaneous updating of shared memory locations. These locations are updated concurrently in a subsequent step. With this in mind, the algorithms for parallel optimization of framed structures are summarized in Tables 17–20. The partitioning of the structures is done according to the algorithm described in Chapter III, while the static analysis phase is performed using the parallel algorithms presented in Chapter IV.

The redesign phase succeeds the analysis phase. Some of the analysis steps are re-visited, while other new steps are introduced. The major steps of this phase are also shown in Fig. 72 and outlined in Table 17. First, the objective function is evaluated. Since this step is computationally inexpensive, it is performed sequentially. Next, the maximum displacement ratio is determined and the active displacement constraints are located. The balance of nodes is used to accomplish the workload balance in this step. The algorithm that performs this step concurrently is summarized in Table 18. The maximum stress ratio for each design variable (cross-sectional area) is then determined. The balance of elements achieved at the end of the initial partitioning stage is used in this step. Subsequently, the maximum stress ratio for the current iteration is found. The concurrent algorithm for this step is summarized in Table 19.

Once the active displacement constraints have been determined, unit loads are applied in the directions of these displacement constraints. For evaluating the gradients of these displacement constraints with respect to the design variables as in eqn (32), the following five steps are re-visited a number of times equal to the number of active

TABLE 17
Algorithms for parallel optimization of framed structures under static loading

1. Read in the input data. Choose an initial design (different cross-sectional areas). Choose ζ_{start} (say, 0·1) and ζ_{end} (say, 0·001). Let $\zeta = \zeta_{start}$.
2. Partition the structure using the three-stage algorithm described in Chapter III:
   (a) Initial partitioning: balance the elements.
   (b) Intermediate partitioning: maximize the number of internal nodes.
   (c) Final partitioning: balance the internal nodes.
   Set *iteration* = 0.
3. Increase the iteration counter by 1 (*iteration* = *iteration* + 1). Set *phase* = 1. Assemble the structure stiffness matrix:
   For *thread* = 1, ... , *num_threads*, do concurrently
   (a) Calculate element stiffness matrices.
   (b) Assemble element stiffness matrices into the structure stiffness matrix (synchronization is required).
4. Assemble the load vector(s):
   For *thread* = 1, ... , *num_threads*, do concurrently
   (s) Assemble element load vectors due to initial stresses (synchronization is required).
   (b) Assemble load vector(s) due to point loads acting at the nodes.
5. Apply the boundary conditions:
   For *thread* = 1, ... , *num_threads*, do concurrently
   (a) Update the load vector(s) (synchronization is required).
   (b) Update the stiffness matrix.
6. Condense the subdomain matrices:
   For *thread* = 1, ... , *num_threads*, do concurrently
   (a) Factorize the subdomain matrices.
   (b) Reduce the subdomain load vectors, and the subdomain stiffness matrices coupling the subdomains with the interface.
   (c) Update the interface load vector and interface stiffness matrix (synchronization is required).
7. Solve for the interface nodal displacements:
   For *thread* = 1, ... , *num_threads*, do concurrently
   Transform the interface stiffness matrix to an upper-triangular matrix.
   For *thread* = 1, ... , *num_threads*, do concurrently
   Adjust the interface load vector through forward substitution.
   For *thread* = 1, ... , *num_threads*, do concurrently
   Find the interface displacements through backward substitution.
8. Retrieve the non-interface displacements:
   For *thread* = 1, ... , *num_threads*, do concurrently
   (a) Reduce the subdomain load vectors and reduce the subdomain stiffness matrices coupling the subdomains with the interface.
   (b) Retrieve the non-interface displacements.
9. Find element forces and stresses (*phase* = 1) or gradients of the active constraint(s) with respect to the design variables (*phase* = 2):

TABLE 17—contd.

For thread = 1, ..., num_threads, do concurrently
(a) Evaluate the elements stiffness matrices.
(b) If phase = 1, evaluate the element forces and stresses; go to step 10. Otherwise, for phase = 2, evaluate the gradients of the active constraint(s) with respect to the design variables according to eqn (32) (synchronization is required); go to step 16.

10. Evaluate the objective function. If this is the last iteration, print the results, and **STOP**. Otherwise, continue to step 11.
11. Set phase = 2. If no displacements are constrained, set the maximum displacement ratio to 1; go to step 12. Otherwise, find the maximum displacement ratio Δ_r, and locate the active displacement constraint(s) according to the algorithm given in Table 18; continue to step 12.
12. Determine the maximum stress ratio for each design variable and the maximum stress ratio for the current iteration Δ_σ according to the algorithm given in Table 19. If no displacements are constrained, go to step 17. Otherwise, continue to step 13.
13. Determine the scaling factor Δ according to the following rule. If $\Delta_\sigma \leq 1$, set $\Delta = \Delta_r$. Otherwise, set $\Delta = \Delta_r \Delta_\sigma$. In subsequent steps, the cross-sectional areas are scaled by a factor of Δ while the displacements and stresses are scaled by a factor of $1/\Delta$.
14. If this is the first iteration, go to step 15. Otherwise, if the value of the objective function for the current iteration is smaller than its value in the previous iteration, continue to step 15. Otherwise, set $\zeta = \frac{1}{2}\zeta$. If $\zeta < \zeta_{end}$, set $\zeta = \zeta_{end}$. Continue to step 15.
15. Apply a unit load in the direction of the most active displacement constraint; go to step 5. This step is repeated for all the active displacement constraints.
16. Calculate the Lagrange multipliers according to eqn (35) or eqn (39).
17. Find the new design variables according to eqn (38), eqn (40), or eqn (41). The concurrent algorithm for finding the new vector of design variables is summarized in Table 20. Go to step 3 to resume the next iteration.

---

displacement constraints: applying the boundary conditions, static condensation, solution of the interface linear equations, retrieving the non-interface displacements, and evaluation of the element forces and stresses. The Lagrange multipliers for the active displacement constraints are determined using eqn (35) or eqn (39). Since this step is computationally inexpensive, it is performed sequentially. Finally, the new design vector is found using eqns (38), (40), or (41). The concurrent algorithm for finding the new vector of design variables is shown in Table 20.

An important feature of this algorithmic procedure is that it is built

TABLE 18
Parallel algorithm for evalatuing the maximum displacement ratio

1. Distribute the number of nodes as evenly as possible among a number of sets equal to the number of prescribed processors.
2. Create a number of threads equal to the number of prescribed processors. Each set is mapped onto a thread.
3. For *threads* = 1, ... , *num_threads*, do concurrently
    For *node* = 1, ... , *num_assigned_nodes*, do sequentially
    For *load* = 1, ... , *num_load_cases*, do sequentially
    Locate the maximum displacement ratio with respect to the thread.
    Next *load*.
    Next *node*.
4. For *threads* = 1, ... , *num_threads*, do sequentially
    Locate the maximum displacement ratio for the iteration
5. For *thread* = 1, ... , *num_threads*, do concurrently
    For *node* = 1, ... , *num_assigned_nodes*, do sequentially
    For *load* = 1, ... , *num_load_cases*, do sequentially
    Locate the active displacement constraints with respect to the thread.
    Next *load*.
    Next *node*.

TABLE 19
Parallel algorithm for evaluating the maximum stress ratio

1. Distribute the number of elements as evenly as possible among a number of sets equal to the number of prescribed processors. The workload balance achieved at the end of the initial partitioning stage is used for acccomplishing this balance.
2. Create a number of threads equal to the number of prescribed processors. Each set is mapped onto a thread.
3. For *thread* = 1, ... , *num_threads*, do concurrently
    For *element* = 1, ... , *num_assigned_elements*, do sequentially
    For *load* = 1, ... , *num_load_cases*, do sequentially
    Locate the maximum stress ratio for each design variable with respect to the thread.
    Next *load*.
    Next *element*.
4. Distribute the number of design variables (cross-sectional areas) as evenly as possible among a number of sets equal to the number of prescribed processors.
5. Create a number of threads equal to the number of prescribed processors. Each set is mapped onto a thread.
6. For *thread* = 1, ... , *num_threads*, do concurrently
    For *variable* = 1, ... , *num_assigned_variables*, do sequentially
    Locate the maximum stress ratio with respect to the design variable.
    Next *variable*.
    Locate the maximum stress ratio with respect to the thread.
7. For *thread* = 1, ... , *num_threads*, do sequentially
    Locate the maximum stress ratio for the iteration.

TABLE 20
Parallel algorithm for evaluating the new vector of design variables

---

1. Distribute the number of design variables (cross-sectional areas) as evenly as possible among a number of sets equal to the number of prescribed processors.
2. Create a number of threads equal to the number of prescribed processors. Each set is mapped onto a thread.
3. For *thread* = 1, ... , *num_threads*, do concurrently
    For *variable* = 1, ... , *num_assigned_variables*, do sequentially
    (a) If the maximum stress ratio for the design variable is greater than or equal to 1 or if the number of constrained displacements is equal to 0, evaluate the new design variable using eqn (41). Otherwise, use eqn (38) or eqn (40).
    (b) If the new design variable is less than the lower bound or greater than the upper bound, set it equal to the lower bound or upper bound, respectively.
    Next *variable*.

---

mostly around the steps involved in the analysis phase. In addition, there is no need to solve an extra set of linear equations for the Lagrange multipliers. For example, consider the case of eqn (38), when the optimal design is based only on the most active displacement constraint. In this case, the redesign in each iteration is obtained mainly at the cost of re-visiting the aforementioned five analysis steps only once. And, even then, major portions of these steps need not be repeated. In fact, all of the operations dealing with the structure stiffness matrix do not have to be repeated. The additional operations are mainly limited to finding the displacements and forces in the structure due to the unit load vector. This makes the parallel algorithms computationally efficient and inexpensive, and particularly suitable for large structures.

Bearing in mind the special storage schemes used for the analysis phase (see Section 4.3) and noting that only one additional load vector at a time is generally required in the redesign phase, the algorithms presented in this chapter can handle large structures at relatively low storage requirements. Moving on to the case of more than one active displacement constraint, we note that the computational cost of each iteration increases with the number of active displacement constraints considered. However, the case where only the most active displacement constraint is considered yields the optimal design without difficulty for a wide range of practical problems, as demonstrated in Section 5.4.

5.4 APPLICATIONS

Several truss and frame problems have been optimized for a variable number of processors on an Encore Multimax shared-memory multiprocessor computer. A minimum of nine elements per subdomain is prescribed as a pre-condition for the initial partitioning stage where a balance of elements is achieved among the subdomains. Similarly, a minimum of three internal nodes per subdomain is prescribed as a pre-condition for the final partitioning stage where a balance of internal nodes is achieved among the subdomains. When any of the pre-conditions is not satisfied, execution is automatically terminated.

Four examples are presented in this chapter: The 200-bar plane truss (Fig. 45), the geodesic dome space truss (Fig. 42), the 90-element braced frame (Fig. 39), and the 798-element 60-storey irregular frame (Fig. 73). In all examples, the case of the most active displacement constraint is first considered, and then the effect of including the set of most active constraints is investigated. In order to assess the convergence properties and stability of the algorithms, the program is allowed to run for 20 iterations for all examples considered in this chapter.

We note that the computational time spent by the sequential code tends to be dominated by the steps dealing with the solution of the entire set of linear equations for the nodal displacements, namely, static condensation, solution of interface linear equations, and retrieving the non-interface displacements (e.g. 78% for the 200-bar plane truss), and then in the steps of calculating the elements forces and stresses and evaluating the gradients (e.g. 12% for the 200-bar plane truss). The time spent on executing the remaining steps such as assembling the stiffness matrix, assembling the load vector, and applying the boundary conditions is relatively small (e.g. 8% for the 200-bar plane truss). The remaining steps of evaluating the maximum displacement, maximum stress ratio, and new design vector are the least significant in terms of execution time (e.g. 2% for the 200-bar plane truss).

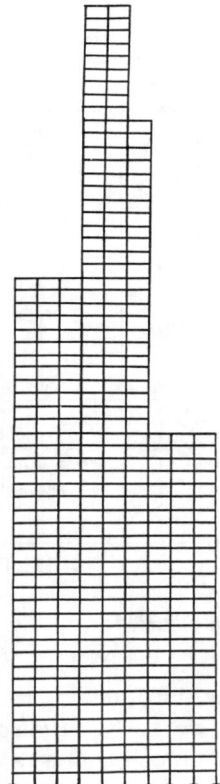

Fig. 73. The 798-element frame.

5.4.1 Optimal design

The 200-bar plane truss shown in Fig. 45 has been studied by Vankayya et al. [1969] and Khan et al. [1979]. All members are made of steel with $E = 30 \times 10^6$ psi and $\rho = 0 \cdot 283 \, \text{lb/in}^3$. Maximum allowable stresses are $\pm 10\,000$ psi. The minimum cross-sectional area is $0 \cdot 1 \, \text{in}^2$. The structure is symmetric about the vertical centerline. This reduces the number of design variables to 105. Three loading cases are considered. First, the structure is subjected to a set of horizontal loads with values of 1 kip acting from left to right at the nodes along the left side of the structure. Second, a set of vertical loads with values of 10 kips acting downwards at the nodes along the vertical lines 1–71, 2–76,

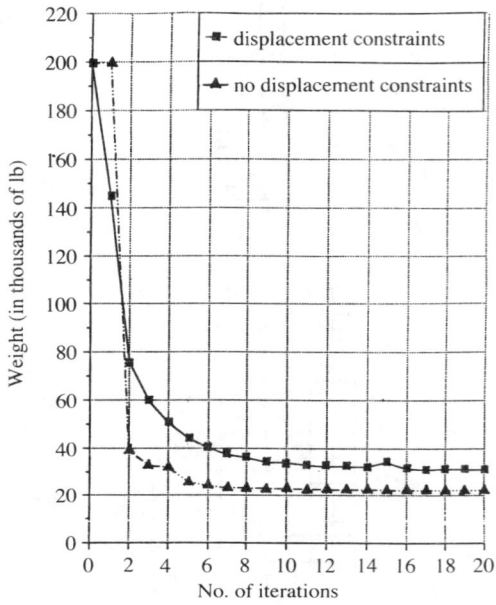

Fig. 74. Iteration history for the 200-bar plane truss.

3–73, 4–77, and 5–75. The third loading case is a combination of the loading cases 1 and 2. Two cases of displacement constraints are considered: no displacement constraints and displacement limits of ±0·5 in imposed on all nodes in the x and y directions. Figure 74 shows the iteration history for both cases. A minimum weight of 21 999 lb is achieved for the case of no displacement constraints, while a minimum weight of 30 696 lb is achieved for the case of constrained displacements, considering the most active displacement constraint only. When the set of most active displacement constraints are considered in the latter case, no improvement in the optimal weight is achieved. Venkayya et al. [1969] reported a minimum weight of 21 116 lb for the case of no displacement constraints and 31 020 lb for the case of constrained displacements. Khan et al. [1979] reported a minimum weight of 32 996 lb for the case of constrained displacements; they did not consider the case with no displacement constraints.

The optimal (minimum weight) design of the geodesic dome space truss shown in Fig. 42 has been studied by Venkayya et al. [1969]. The modulus of elasticity is 10 000 ksi, while the specific weight of the truss material is 0·1 lb/in^3. Allowable stresses are limited to ±25 000 psi. A minimum cross-sectional area of 0·1 in^2 is used. Displacement limits of ±0·1 in are imposed on all nodes in the x, y, and z directions. The structure is symmetric about the line connecting nodes 5 and 57 and that connecting nodes 12 and 50. This reduces the number of design variables to 36. Four loading cases are considered, in all of which all loads are unit loads (1 kip) acting downwards in the z direction at the nodes. In the first case, node 31 is the only loaded node. In the second case, all of the nodes on the line joining the nodes 5 and 57 and those nodes to the left of that line are loaded. In the third case, all of the nodes on the line joining the nodes 5 and 57 and those nodes to the right of that line are loaded. In the fourth case, all of the nodes are loaded. An optimal design of 200·4 lb is obtained when only the most active displacement constraint is considered. When the set of most active displacement constraints is considered, a minimum weight of 184·3 lb is achieved. Figure 75 shows the iteration history for the latter case. Venkayya et al. [1969] reported a minimum weight of 180·9 lb.

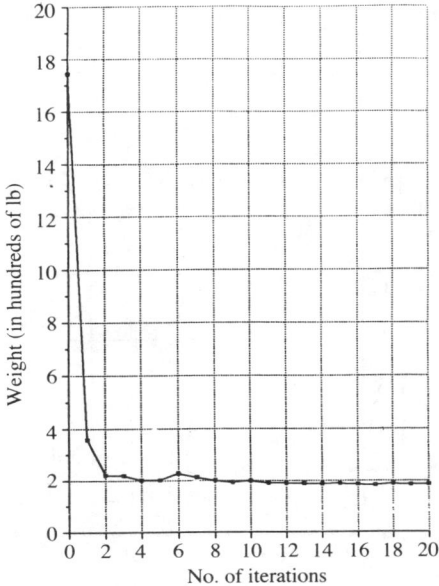

Fig. 75. Iteration history for the geodesic dome space truss.

The 90-element braced frame shown in Fig. 39 is adopted from Cheng et al. [1981]. The modulus of elasticity of the frame material is 29 000 ksi. The specific weight is $0.283 \, lb/in^3$. The allowable stresses in the moment resisting elements are $\pm 29\,000$ psi, while those in the truss elements (bracings) are $\pm 20\,000$ psi. The minimum cross-sectional area is $0.5 \, in^2$. Displacements of the nodes are restricted to $\pm 0.005h$ in the horizontal direction, where h is the elevation of the node above the ground. The structure is symmetric about the vertical centerline. This reduces the number of design variables to 45. Three loading cases are considered. First, the horizontal moment resisting elements are subjected to a uniform distributed load of 180 lb/in, acting downwards along their spans. Second, the nodes at the left side of the structure are subjected to horizontal loads with magnitude of 9 kips acting from left to right. The third loading case is a summation of the first two cases. For the moment resisting elements, the moments of inertia are taken to be 75 times the cross-sectional areas, and the section moduli to be 9 times the cross-sectional areas. These relationships appear to be representative of commonly used wide flange shapes, and are

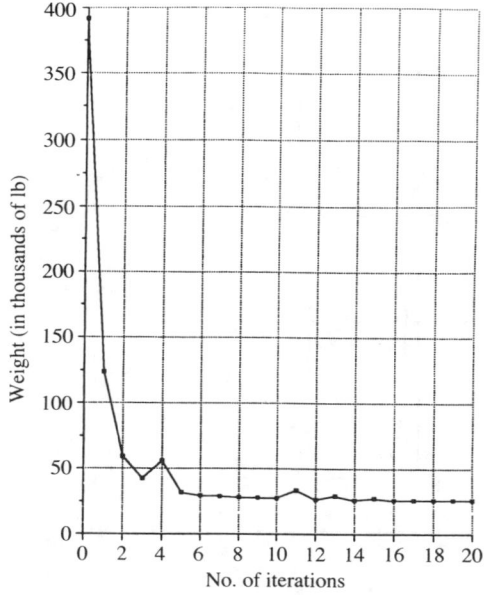

Fig. 76. Iteration history for the 90-element braced frame.

chosen in order to maintain linearity between element stiffness matrices and element cross-sectional areas. Figure 76 shows the iteration history when only the most active displacement constraint is considered. An optimal design of 25 790 lb is achieved. When the set of most active displacement constraints is considered, no improvement in the optimal minimum weight is achieved. The aforementioned loading cases were not considered by Cheng et al. [1981]. Rather, they considered earthquake loading and response-spectrum analysis.

The 798-element frame shown in Fig. 73 is introduced as an example to illustrate the capability of the parallel algorithms in handling optimization of large structures with a large number of design variables and constraints. The girders and the columns are assumed to have constant lengths of 20 and 12 ft, respectively. The modulus of elasticity is 29 000 ksi and the specific weight is 0·283 lb/in^3. The allowable stresses in the elements are ±22 000 psi. The minimum cross-sectional area is 5·0 in^2. Displacements of the nodes are restricted to ±0·005h in the horizontal direction, where h is the elevation of the node from ground level. The girders in each storey are represented by one variable and the columns by another variable. Thus, the number of design variables is 120. Three loading cases are considered. First, the girders are subjected to a uniform distributed load of 120 lb/in acting downwards along their spans. Second, the nodes at the left side of the structure are subjected to horizontal loads with magnitude of 0·3n kips acting from left to right, where n is the storey number measured from the ground. The third loading case is the combination of the first two cases. The relationships between the moments of inertia and the cross-sectional areas and the section moduli are the same as those described in the previous example. Figure 77 shows the iteration history when only the most active displacment constraint is considered. An optimal design of 2401 kips is achieved. Including the set of most active displacement constraints does not change the results appreciably.

At this point, we note that the case when only the most active displacement constraint is included in the formulation yields the optimal design without difficulty. Even for the worst case of the geodesic dome space truss, a 'near optimum' weight was obtained when only the most active displacement constraint was included. For all the examples, the local minimum was obtained after 5–9 iterations. These results are representative of the algorithms' performance. Another feature of the algorithms in this case is that the workload in

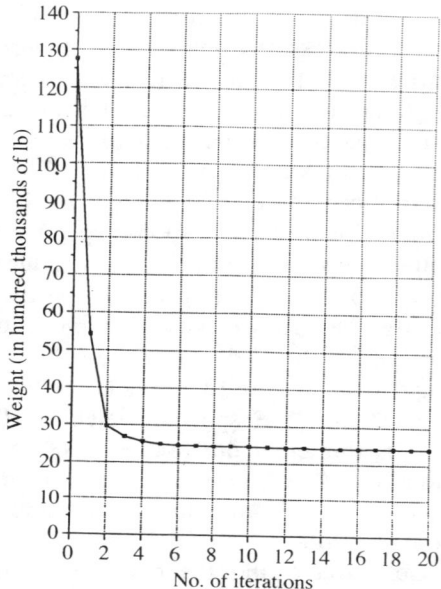

Fig. 77. Iteration history for the 798-element frame.

each iteration is about the same. Consequently, the performance of the algorithm can be assessed through the study of a single iteration. A single optimization iteration consists of two phases: analysis and redesign. In the following, the concurrent performance of the computational steps in a single iteration is discussed for the 200-bar plane truss. The overall performance within an iteration is also presented for the four examples described earlier.

5.4.2 Assembly and application of boundary conditions

The steps of assembling the structure stiffness matrix, load vector(s), and applying the boundary conditions are visited during the analysis phase, while only the step of applying the boundary conditions is re-visited during the redesign phase. Speed-ups for each of the three steps for the 200-bar plane truss are shown in Fig. 78. When the overhead time required for creating the threads is neglected, the speed-up curve is practically linear (the workload balance curve). The larger the ratio of the time required for creating the threads to

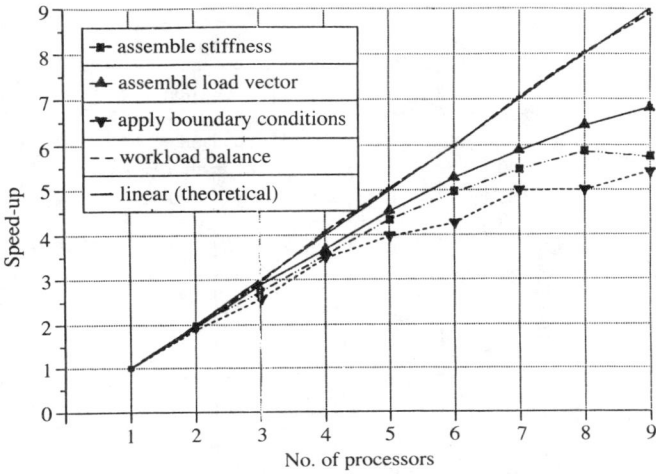

Fig. 78. Speed-ups for assembling the structure stiffness matrix, load vector, and applying the boundary conditions for the 200-bar plane truss.

accomplish a certain step to the sequential time required for executing the step, the more is there an offset from the theoretical linear speed-up.

5.4.3 Static condensation

In this step, the effect of the linear equations associated with the non-interface degrees of freedom is collapsed onto those associated with the interface. Three basic stages are involved: factorizing the subdomain stiffness matrices, reducing the subdomain load vectors and the subdomain stiffness matrices coupling the subdomains with the interface, and updating the interface loading vector and the interface stiffness matrix. All stages are visited during the analysis phase, while only reduction of the subdomain load vectors and updating of the interface load vector are revisited during the redesign phase. The times spent by the sequential code on executing each of the three stages are shown in Fig. 79 as percentages of the total sequential time required for the static condensation step for the 200-bar plane truss for a variable number of processors. As the number of processors increases, the execution time for 'factorization' tends to be less significant, while the remaining execution time is shared between

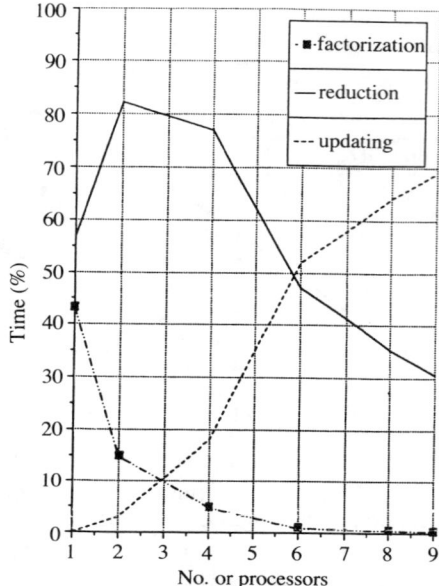

Fig. 79. The times spent by the sequential code on executing each of the three stages of static condensation as percentages of the total sequential time required for the static condensation for the 200-bar plane truss.

'reduction' and 'updating' with the gradual dominance of 'updating'. This is due to the fact that the number of interface nodes increases as the number of subdomains (and consequently processors) increases. Figure 80 shows the number of internal nodes assigned to each subdomain and the number of interface nodes for the 200-bar plane truss. Figure 81 compares the speed-ups attained for the cases of using threads and the workload balance.

5.4.4 Solution of interface linear equations

The linear equations associated with the interface degrees of freedom are solved in this step, which consists of three stages: transforming the matrix to an upper-triangular one, and forward and backward substitutions. Each stage requires a number of sets of threads equal to 3 times the *number of interface nodes*. All stages are visited during the analysis phase, while only the stages of forward and backward

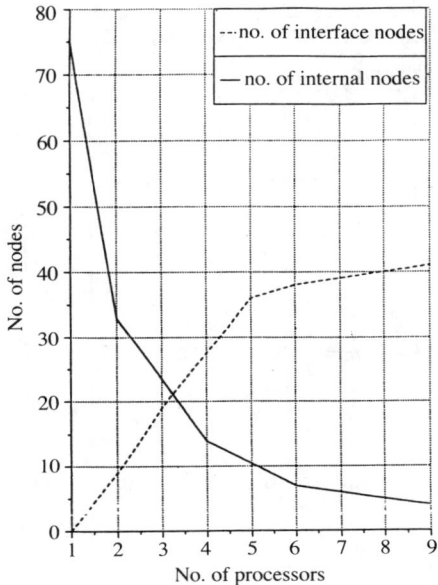

Fig. 80. Numbers of interface and internal nodes for the 200-bar plane truss for various numbers of processors in use.

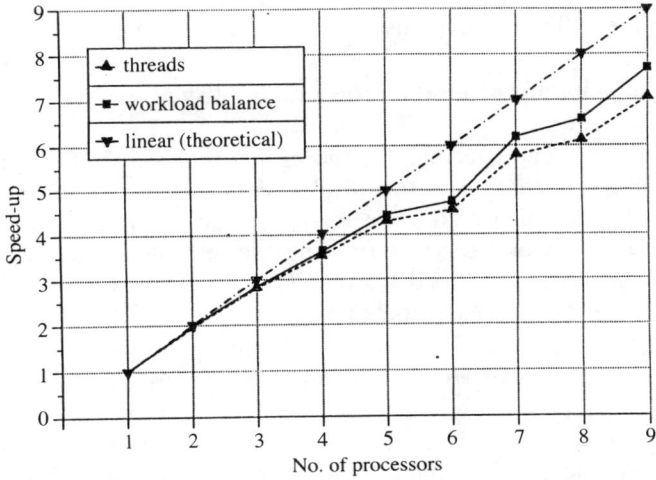

Fig. 81. Speed-ups for the static condensation step for the 200-bar plane truss.

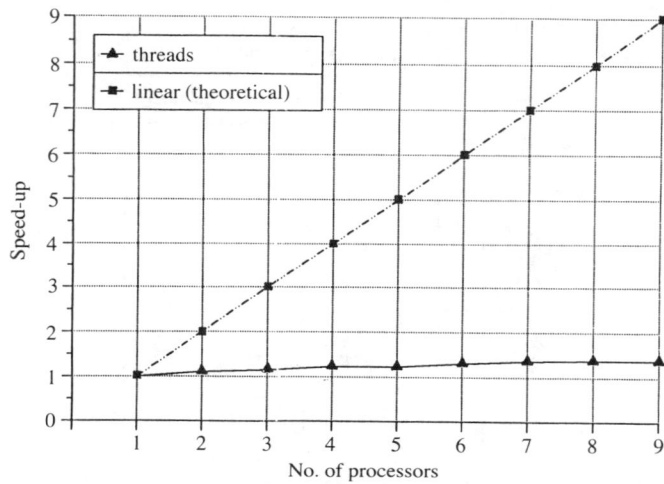

Fig. 82. Speed-ups for the solution of interface linear equations for the 200-bar plane truss.

substitutions are re-visited during the redesign phase. Since the overhead time required for creating the threads is added sequentially to the execution time of the concurrent application, its effect on the speed-ups attained for this step is severe. Figure 82 shows the speed-ups achieved during the solution of the interface linear equations for the 200-bar plane truss.

5.4.5 Computation of non-interface nodal displacements

Once the displacements corresponding to the interface degrees of freedom have been found, the displacements corresponding to non-interface degrees of freedom are retrieved. This step is visited during the analysis phase and re-visited during the redesign phase. Figure 83 shows the speed-ups attained for the 200-bar plane truss for the cases of threads and the workload balance.

5.4.6 Computation of element forces and stresses

This step is visited during the analysis and re-visited during the redesign. In the analysis phase, the element forces and stresses are evaluated. The gradients of the most active displacement constraint

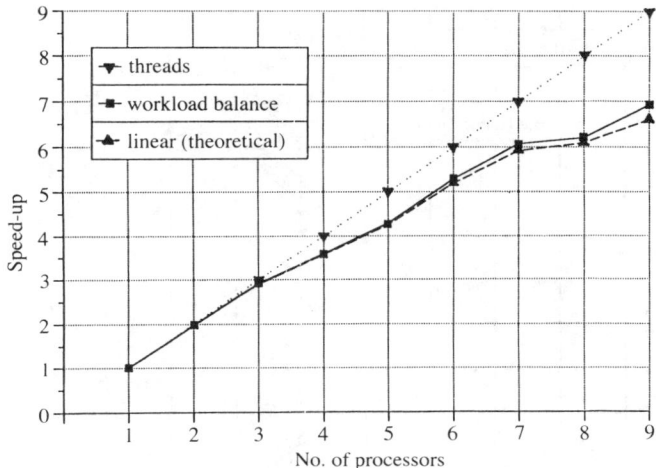

Fig. 83. Speed-ups for the computation of non-interface displacements for the 200-bar plane truss.

with respect to the design variables are calculated during the redesign phase. The speed-ups for the cases of threads and workload balance for the 200-bar plane truss are shown in Fig. 84. The speed-up for the case of workload balance is practically linear, indicating that the computations for the forces and stresses (and gradients) are identical for all the elements.

5.4.7 Overall speed-up, workload balance, and efficiency

The overall performance of the parallel optimization algorithms is now assessed. To demonstrate the performance of the algorithms presented in this chapter, comparisons of the overall speed-up, workload balance, and efficiency are included. A comparison of the overall speed-ups for a single iteration for the 200-bar plane truss is shown in Fig. 85. The effect of the number of threads created during the step of solving the interface linear equations on reducing the overall speed-ups is evident. Also, note that better speed-ups are achieved for the entire optimization procedure (both phases) compared with the analysis phase only. This indicates that a better workload balance is achieved during the redesign phase than the analysis phase. Figure 86 shows the speed-ups attained during a single iteration of the entire

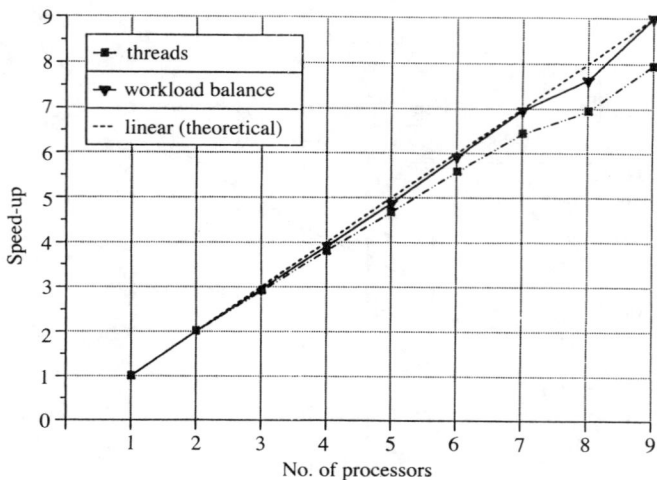

Fig. 84. Speed-ups for the computation of forces and stresses (and gradients) for the 200-bar plane truss.

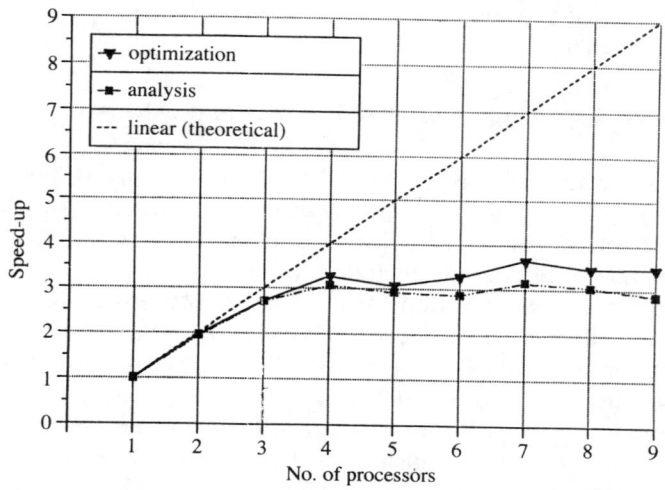

Fig. 85. Overall speed-ups for the 200-bar plane truss.

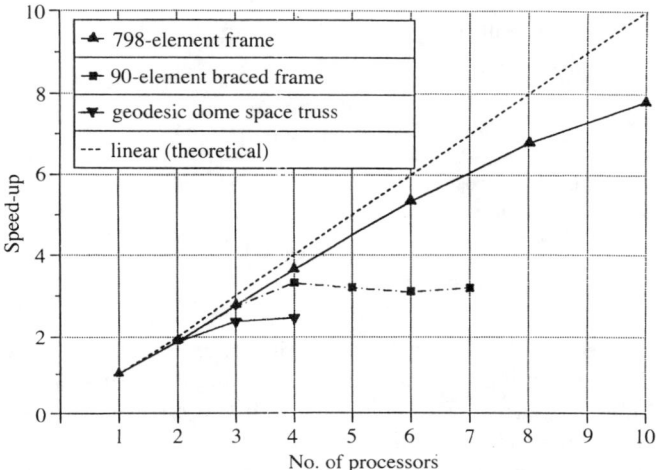

Fig. 86. Overall speed-ups in the optimization process for the geodesic dome space truss, the 90-element braced frame, and the 798-element frame.

optimization procedure (both phases) for the geodesic dome space truss, the 90-element braced frame, and the 798-plane frame. Execution is automatically terminated when the minimum number of elements (nine) or internal nodes (three) is not satisfied within the subdomains. This represents the case of ten or more subdomains for the 200-bar plane truss, five or more subdomains for the geodesic dome space truss, and eight or more subdomains for the 90-element plane frame. Better speed-ups are achieved when the percentage of the sequential time required for the step of solving the interface linear equations is small compared with the total sequential time and/or the overhead time required for thread creation in this step is small compared with the time required for accomplishing this step.

Table 21 lists the speed-up for the four examples considered for the cases of 4, 7, 9, and 10 processors. The numbers of degrees of freedom for each structure are also indicated. The results show that better speed-ups are achieved for larger structures when the same number of processors is used. On one hand, this is due to the fact that the overhead time required for creating a thread tends to be less significant compared with the time required for the thread to execute its task as the size of the task assigned to each thread increases. On the other hand, the percentage of the time consumed in the fine-grained

TABLE 21
Overall speed-up for the example problems

Example	Number of degrees of freedom	Number of processors			
		4	7	9	10
Geodesic dome space truss	111	2·49	—	—	—
90-element braced frame	135	3·31	3·2	—	—
200-bar plane truss	150	3·33	3·66	3·48	—
798-element frame	1287	3·64	6·06	7·29	7·80

parallelism step (solution of the linear interface equations) tends to be less significant compared with the overall execution time as the size of the problem increases. This explains why the speed-up curve for the 798-element frame is higher than the 90-element frame curve, which in turn is higher than the geodesic dome space truss curve in Fig. 86. When the overhead time required for creating the threads and the time required for solving the set of linear interface equations become more significant relative to the execution time, the concurrent application fails to pick up additional speed-ups as the number of processors increases. This is shown in Fig. 85 for the 200-bar plane truss for the cases of more than seven processors and in Fig. 86 for the 90-element plane frame for the case of more than four processors.

The overall workload balance (ignoring the overhead time required for thread creation) for the 200-bar plane truss is shown in Fig. 87 for two cases: analysis phase only and optimization (both phases). Figure 88 shows the overall workload balance for a single iteration of the optimization process for the geodesic dome space truss, the 90-element braced frame, and the 798-plane frame. The offsets of the curves in Figs 87 and 88 from the theoretical linear speed-up case are mainly due to imbalance of workload. The effect of this imbalance becomes more significant as the number of threads (and processors) increases and consequently the task of each thread becomes smaller. Similarly, the overall efficiency for the analysis (first phase only) and the optimization (both phases) for the 200-bar plane truss is shown in Fig. 89. Figure 90 shows the overall efficiency in the optimization process for the geodesic dome space structure, the 90-element braced frame, and the 798-element plane frame. For the largest example, the 798-element frame, an overall efficiency of 85% is achieved when ten processors are in use.

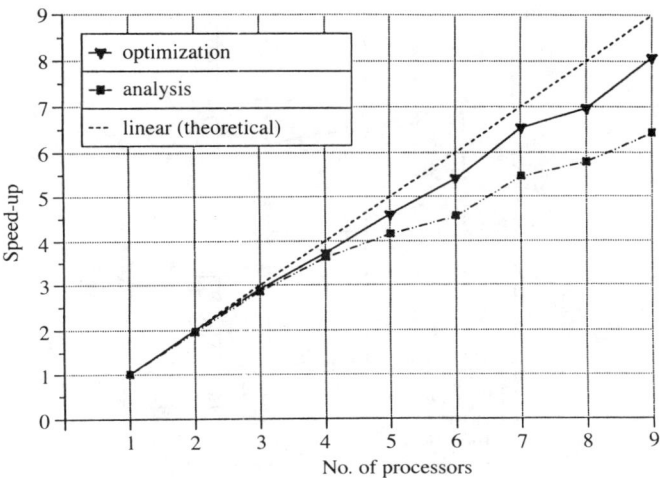

Fig. 87. Overall workload balance for the 200-bar plane truss.

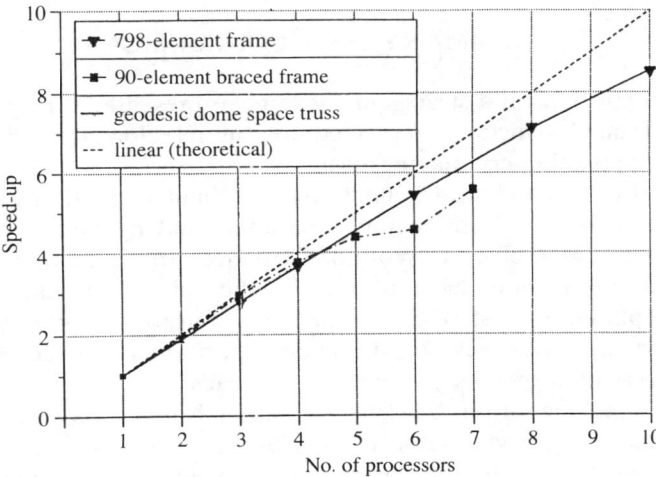

Fig. 88. Overall workload balance in the optimization process for the geodesic dome space truss, the 90-element braced frame, and the 798-element frame.

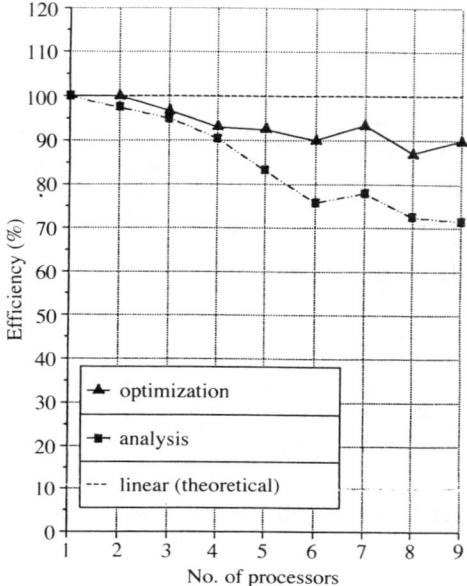

Fig. 89. Overall efficiency for the 200-bar plane truss.

5.5 SUMMARY AND CONCLUSIONS

Parallel algorithms and stratagems have been presented for optimization of framed structures subjected to multiple loading conditions using the optimality criteria approach.

The algorithms and stratagems have been implemented in C on an Encore Multimax shared-memory computer, and applied to several truss and frame problems. Workload balance stratagems have been presented for maintaining a balance in the calculations during the redesign phase. The step of solving the interface linear equations represents the bottleneck for the overall performance owing to the need for creation of a large number of threads. For large structures, however, the effect of this bottleneck situation is less severe and better performance is achieved, since the percentage of the sequential time required for solving the interface linear equations is small compared with the overall sequential time. This shows the merit of the overall approach presented in this book in reducing the number of linear

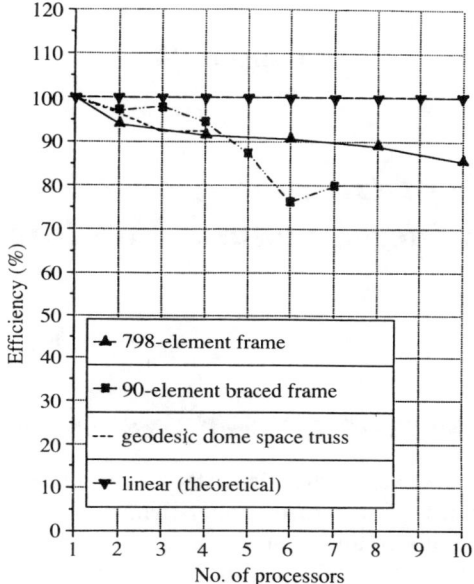

Fig. 90. Overall efficiency in the optimization process for the geodesic dome space truss, the 90-element braced frame, and the 798-element frame.

equations required to be solved simultaneously to those only associated with the interface through the concept of substructuring.

The optimization algorithms are built mostly around the steps involved in the analysis phase. In concurrent processing terms, this makes the concurrent optimization algorithms perform as well as (if not better than) the concurrent analysis algorithms. In optimization terms, the efficiency of the algorithms is independent of the number of design variables and behavior constraints to a large extent, thus allowing problems with a large number of design variables and behavior constraints to be handled effectively. The concurrent algorithms are particularly effective for the optimization of large structures, as can clearly be seen from Figs 86 and 88.

Chapter VI

Concurrent Optimization of Structures under Dynamic Loading

6.1 OPTIMIZATION UNDER DYNAMIC LOADING

In this chapter, we extend the algorithms presented in Chapter V to include optimization of structures under dynamic loading. The parallel algorithms are general and can be used for any kind of loading. However, for the sake of illustration, we consider only the case of seismic loading. First, we summarize the equations involved in the analysis of structures subjected to combined static and dynamic seismic loadings. Assuming that the structure has been discretized into a finite number of elements, the dynamic equilibrium equation (or equation of motion) can be written as

$$\mathbf{M}(\mathbf{x})\ddot{\mathbf{u}} + \mathbf{C}(\mathbf{x})\dot{\mathbf{u}} + \mathbf{K}(\mathbf{x})(\mathbf{u} + \mathbf{r}) = \mathbf{D} + \mathbf{R} \qquad (42)$$

where **M**, **C**, and **K** are the structure mass, damping, and stiffness matrices, respectively. The size of these matrices is $k \times k$, where k corresponds to the number of degrees of freedom of the structure. The vectors **u** and **r** represent the dynamic and static displacements of the structure, respectively. The vectors **D** and **R** denote the applied dynamic and static loads, respectively. An overdot represents differentiation with respect to the time variable. The elements of the matrices **M**, **C**, **K**, and **r** are functions of the vector **x**, the components of which are the V design variables $x_1, x_2, \ldots, x_V$ defining the structure. The vectors $\ddot{\mathbf{u}}$, $\dot{\mathbf{u}}$, **u**, and **D** are functions of both the design variables and the time parameter t. The components of **R** are constants. Equation (42) can be decomposed into separate equilibrium equations: one corresponding to the static loading and the other corresponding to the dynamic loading. Noting that $D = F(x)H(t)$, where $F(x)$ represents the spatial distribution of the seismic load and $H(t)$ is the time-dependent function of a specified horizontal ground acceleration, we can write

$$K(x)r = R \qquad \text{(static case)} \qquad (43)$$

$$M(x)\ddot{u} + C(x)\dot{u} + K(x)u = F(x)H(t) \qquad \text{(dynamic case)} \qquad (44)$$

Assuming linear elastic behavior, each analysis problem may be addressed separately, and the complete solution may be obtained by superposition. For a given vector of design variables, x, the stiffness matrix K can be evaluated and the static displacements vector r can be obtained through the solution of eqn (43). To decouple eqn (44), the following transformation is introduced:

$$u = \Phi z \qquad (45)$$

where Φ is a $k \times n$ matrix whose columns are the n eigenmodes of the structure $(n < k)$ and z is the generalized coordinate vector or reduced state variables vector of dimension n. The modal shape matrix Φ results from solving the eigenproblem $M\Phi\Lambda - K\Phi = 0$, where Λ is an $n \times n$ diagonal matrix of the eigenvalues. Substituting eqn (45) into eqn (44) and pre-multiplying by Φ^T, we obtain

$$\Phi^T M \Phi \ddot{z} + \Phi^T C \Phi \dot{z} + \Phi^T K \Phi z = \Phi^T D$$

or

$$M^* \ddot{z} + C^* \dot{z} + K^* z = D^* \qquad (46)$$

where M^*, C^*, and K^* are $n \times n$ diagonal matrices, and z, $\dot{z}$, $\ddot{z}$, and D^* are column vectors with n rows. Thus, the ith uncoupled mode can be represented as

$$\ddot{z}_i + 2c_i\omega_i\dot{z}_i + \omega_i^2 z_i = D_i^* \qquad (47)$$

where ω_i and c_i are the natural frequency and the damping ratio of the mode i, respectively, and

$$2c_i\omega_i = \frac{\varphi_i^T C \varphi_i}{\varphi_i^T M \varphi_i}, \qquad \omega_i^2 = \frac{\varphi_i^T K \varphi_i}{\varphi_i^T M \varphi_i} \qquad (48)$$

The solution of eqn (47) is referred to as the temporal solution. It can be expressed in terms of the following Duhamel integral (assuming $\omega_i' = \omega_i\sqrt{1 - c_i^2}$):

$$z_i(t) = \frac{1}{\omega_i'} \int_0^t P_i^*(\tau) e^{-c_i\omega_i(t-\tau)} \sin \omega_i'(t - \tau) \, d\tau \qquad (49)$$

However, we use a direct integration method, that is the Wilson-Θ method (Bathe [1982]) for finding the values of z.

The substitution of the values of z into eqn (45) yields the dynamic displacement vector **u**. The total displacement vector **d** resulting from the combined static and dynamic loadings can be computed as follows:

$$\mathbf{d} = \mathbf{r(x)} \pm \mathbf{u(x}, t) \qquad (50)$$

The ± sign indicates the fact that the dynamic seismic loading can be in either direction. The vector of internal forces for the ith member, $\mathbf{S}_i$, can be determined by

$$\mathbf{S}_i = \mathbf{K}_i\mathbf{v}_i = \mathbf{K}_i\mathbf{a}_i\mathbf{d} \qquad (51)$$

where $\mathbf{v}_i$ is the displacement vector for the ith element, $\mathbf{a}_i$ is the compatibility matrix relating the displacements of the ith member to those of the structural system, and $\mathbf{K}_i$ is the element stiffness matrix for the ith member. Finally, the vector of stresses for the member i, $\mathbf{\sigma}_i$, can be obtained from the relation

$$\mathbf{\sigma}_i = \mathbf{B}_i\mathbf{S}_i = \mathbf{B}_i\mathbf{K}_i\mathbf{a}_i\mathbf{d} \qquad (52)$$

where $\mathbf{B}_i$ is the stress transformation matrix containing the cross-sectional properties of the ith member.

Now, the optimum structural design problem under the combined static and dynamic loadings can be cast in the following form: find the vector of design variables **x** such that

$$W(\mathbf{x}) = \sum_{m=1}^{M} \rho_m l_m A_m(x_m) \to \min \qquad (53)$$

subject to

$$d_p^l \leq r_{pg}(\mathbf{x}) \pm u_p(\mathbf{x}, t) \leq d_p^u \quad (p = 1, 2, \ldots, P; g = 1, 2, \ldots, G) \qquad (54)$$

$$\sigma_m^l \leq \sigma_{mg}(\mathbf{x}) \pm \sigma_m(\mathbf{x}, t) \leq \sigma_m^u$$
$$(m = 1, 2, \ldots, M; g = 1, 2, \ldots, G) \qquad (55)$$

$$x_v^l \leq x_v \leq x_v^u \quad (v = 1, 2, \ldots, V) \qquad (56)$$

where $W(\mathbf{x})$ is the objective function represented by the weight of the structure, ρ_m is the mass density of the mth member, l_m is the length of the mth member, $A_m(x_m)$ is the cross-sectional area of the mth element, and M is the total number of elements. The constants d_p^l, d_p^u, σ_m^l, σ_m^u, x_v^l, and x_v^u are the lower and upper bounds on the nodal displacements, member stresses, and design variables, respectively. The constants P, G, and V represent the number of constrained displacements, static loading cases, and design variables, respectively. The functions r_{pg} and σ_{mg} represent the pth displacement and the stress in the mth member for the gth static loading case, respectively. Similarly, the functions u_p and σ_m represent the pth displacement and the stress in the mth member due to the dynamic loading, respectively.

The time parameter t in eqns (54) and (55) can be eliminated through searching for the maximum displacement and stress for all t ($t > 0$). In this case, eqns (54) and (55) can be written in the form

$$d_p^l \leq d_{pg} \leq d_p^u \quad (p = 1, 2, \ldots, P; g = 1, 2, \ldots, G) \quad (57)$$

$$\sigma_m^l \leq \bar{\sigma}_{mg} \leq \sigma_m^u \quad (m = 1, 2, \ldots, M; g = 1, 2, \ldots, G) \quad (58)$$

where

$$d_{pg} = r_{pg} \pm \max_{t>0} \{|u_p(\mathbf{x}, t)|\}, \quad \bar{\sigma}_{mg} = \sigma_{mg} \pm \max_{t>0} \{|\sigma_m(\mathbf{x}, t)|\}$$

To this end, the rest of the chapter is organized as follows. In Section 6.2, the algorithm for assembling the structure mass matrix concurrently is presented. The parallel solution of the eigenproblem is described in Section 6.3. Then, the concurrent temporal solution is outlined in Section 6.4. The parallel evaluation of the dynamic displacements and stresses is given in Section 6.5. Figure 91 and Table 22 outline the entire concurrent solution process. Applications are described in Section 6.6, and a summary and conclusions are given in Section 6.7. Note that the substructuring algorithm presented in Chapter III and the parallel algorithms for static analysis of structures described in Chapter IV are used for partitioning the structure and for the static analysis, respectively.

6.2 ASSEMBLING THE STRUCTURE MASS MATRIX

The structure mass matrix is assembled in this step. The consistent mass matrix is used in this work. The balance of elements achieved at the end of the initial partitioning stage is used as the workload balance stratagem. The concurrent algorithm is described in Table 23. Each

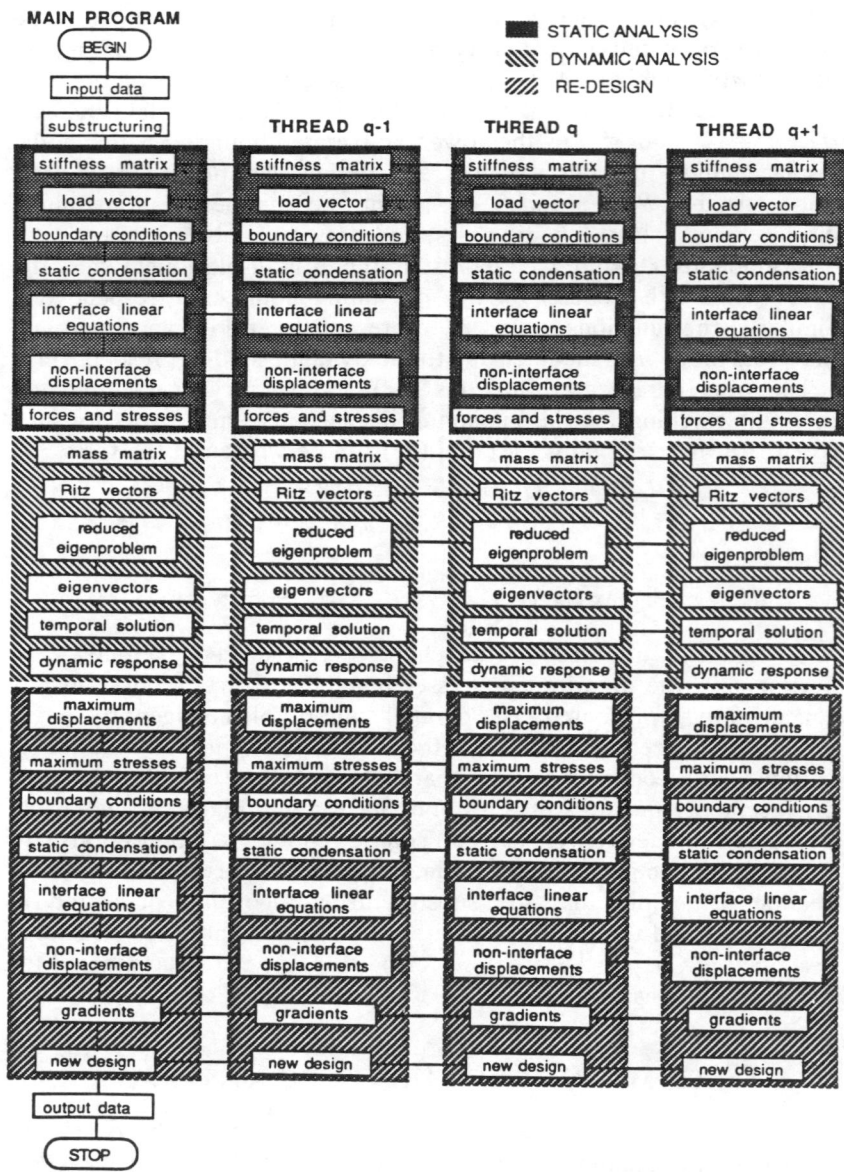

Fig. 91. Macro flow chart of the parallel C program for optimization of framed structures subjected to combined loading.

TABLE 22
Algorithms for parallel optimization of framed structures under combined loadings

1. Read in the input data. Choose an initial design (different cross-sectional areas). Choose ζ_{start} (say, 0·1) and ζ_{end} (say, 0·001). Let $\zeta = \zeta_{start}$.
2. Partition the structure using the three-stage substructuring algorithm described in Chapter III.
   (a) Initial partitioning: balance the elements.
   (b) Intermediate partitioning: maximize the number of internal nodes.
   (c) Final partitioning: balance the internal nodes.
   Set *iteration* = 0.
3. Increase the iteration counter by 1 (*iteration* = *iteration* + 1). Set *phase* = 1. Assemble the structure stiffness matrix:
   For *thread* = 1, . . . , *num_threads*, do concurrently
   (a) Calculate element stiffness matrices.
   (b) Assemble element stiffness matrices into the structure stiffness matrix (synchronization is required).
4. Assemble the load vector(s):
   For *thread* = 1, . . . , *num_threads*, do concurrently
   (a) Assemble elements load vectors due to initial stresses (synchronization is required).
   (b) Assemble load vector(s) due to point loads acting at the nodes.
5. Apply the boundary conditions:
   For *thread* = 1, . . . , *num_threads*, do concurrently
   (a) Update the load vector(s) (synchronization is required).
   (b) Update the stiffness matrix.
6. Condense the subdomain matrices:
   For *thread* = 1, . . . , *num_threads*, do concurrently
   (a) Factorize the subdomain matrices.
   (b) Reduce the subdomain load vectors and reduce the subdomain stiffness matrices coupling the subdomains with the interface.
   (c) Update the interface load vector and interface stiffness matrix (synchronization is required).
7. Solve for the interface nodal displacements:
   For *thread* = 1, . . . , *num_threads*, do concurrently
   Transform the interface stiffness matrix to an upper-triangular matrix.
   For *thread* = 1, . . . , *num_threads*, do concurrently
   Adjust the interface load vector through forward substitution.
   For *thread* = 1, . . . , *num_threads*, do concurrently
   Find the interface displacements through backward substitution.
8. Retrieve the non-interface displacements:
   For *thread* = 1, . . . , *num_threads*, do concurrently
   (a) Reduce the subdomains load vectors and reduce the subdomain stiffness matrices coupling the subdomains with the interface.
   (b) Retrieve the non-interface displacements.

(*continued*)

TABLE 22—*contd*.

9. Find element forces and stresses (or gradients of the active constraint(s) with respect to the design variables):
   For *thread* = 1, . . . , *num_threads*, do concurrently
   (a) Evaluate the element stiffness matrices.
   (b) If *phase* = 1, evaluate the element forces and stresses; go to step 10. Otherwise, if *phase* = 2, evaluate the gradients of the active constraint(s) with respect to design variables according to eqn (59) (synchronization is required); go to step 23.
10. Assemble the structure mass matrix as shown in Table 23:
    For *thread* = 1, . . . , *num_threads*, do concurrently
    (a) Calculate element mass matrices.
    (b) Assemble element mass matrices into the structure mass matrix (synchronization is required).
11. Evaluate the Ritz vectors as shown in Table 24.
12. Solve for the eigenvalues and eigenvectors of the reduced system. The concurrent algorithm is shown in Table 26.
13. Calculate the eigenvalues and eigenvectors (final Ritz vectors) of the original problem. This step involves matrix operations performed concurrently as shown in Table 25.
14. Compute the modal displacements. The concurrent algorithm for the temporal solution is outlined in Table 27.
15. Determine the displacement vector at each time increment. This step involves multiplication of matrices performed concurrently as shown in Table 25.
16. Find elements forces and stresses at each time increment:
    For *thread* = 1, . . . , *num_threads*, do concurrently
    (a) Evaluate the elements stiffness matrices.
    (b) Evaluate the elements forces and stresses.
17. Evaluate the objective function. If this is the last iteration, print the results, and **STOP**. Otherwise, continue to step 18.
18. Set *phase* = 2. If no displacements are constrained, set the maximum displacement ratio to 1; go to step 19. Otherwise, find the maximum displacement ratio Δ_r and locate the active displacements constraint(s) concurrently as presented in Table 18. Continue to step 19.
19. Determine the maximum stress ratio for each design variable and the maximum stress ratio for the current iteration Δ_σ concurrently as shown in Table 19. If no displacements are constrained, go to step 24. Otherwise, continue to step 20.
20. Determine the scaling factor Δ according to the following rules. If $\Delta_\sigma \leq 1$, set $\Delta = \Delta_r$. Otherwise, set $\Delta = \Delta_r \Delta_\sigma$. In subsequent steps, the cross-sectional areas are scaled by a factor of Δ while the displacements and stresses are scaled by a factor of $1/\Delta$.
21. If this is the first iteration, go to step 22. Otherwise, if the value of the objective function for the current iteration is smaller than the corresponding value for the previous iteration, continue to step 22. Otherwise, set $\zeta = \frac{1}{2}\zeta$. If $\zeta < \zeta_{end}$, set $\zeta = \zeta_{end}$. Continue to step 22.

TABLE 22—contd.

22. Apply a unit load in the direction of the most active displacement constraint; go to step 5. This step is repeated for all the active displacement constraints.
23. Calculate the Lagrange multipliers according to eqn (35) or eqn (39).
24. Find the new vector of design variables according to eqn (38), eqn (40), or eqn (41). If the maximum stress ratio for the vector of design variables is greater than or equal to 1, or if the number of constrained displacements is equal to 0, evaluate the new design variables using eqn (41). Otherwise, use eqn (38) or eqn (40). The concurrent algorithm is described in Table 20. Go to step 3 to resume the next iteration.

TABLE 23
Algorithm for assembling the structure mass matrix

1. Distribute the number of elements as evenly as possible among a number of sets equal to the number of prescribed processors. The balance of elements achieved at the end of the initial partitioning stage is used to determine these sets.
2. Create a number of threads equal to the number of prescribed processors. Each set is mapped onto a thread.
3. For $q = 1, \ldots, num\_threads$, do concurrently
    For $elmnt = 1, \ldots, num\_assigned\_elmnts$, do sequentially
    (a) Set up the element mass matrix.
    (b) Assemble the element mass matrix into the structure mass matrix. Avoid a synchronization situation through using extra storage locations for the overlapped locations in zone 4 (see Figs 53 and 54) of the structure mass matrix.
    Next *element*.
4. Distribute this number of the overlapped locations of the interface portion of the structure mass matrix as evenly as possible among a number of sets equal to the number of prescribed processors.
5. Create a number of threads equal to the number of prescribed processors. Each set is mapped onto a thread.
6. For $q = 1, \ldots, num\_threads$, do concurrently
    For $add = 1, \ldots, num\_assigned\_additions$, do sequentially
    (a) Add-up the individual components of the overlapped location.
    (b) Store the sum in its appropriate location of the structure mass matrix.
    Next *add*.

thread calculates the mass matrices for the elements assigned to it and then assembles them into the structure mass matrix. A racing condition is avoided using the stratagem of extra storage outlined in Table 23.

6.3 SOLUTION OF THE EIGENPROBLEM

In this section, we parallelize the Wilson–Ritz algorithm developed by Leger et al. [1986] for use on shared-memory multiprocessor computers. This algorithm is developed for linear dynamic analysis of structures subjected to a fixed spatial distribution of loads such as the case of earthquake ground motions. First, a number of Ritz vectors equal to the number of eigenmodes, n, are generated, satisfying the conditions of orthonormality with respect to the structure mass matrix. The parallel algorithm is presented in Table 24. In addition to the solution of linear equations (step 1(b) of Table 24), whose parallelization has already been outlined in Table 22 and shown in Fig. 91, Table 24 involves other steps that need to be parallelized. These are matrix operations such as matrix multiplications and inner products. The algorithms that perform parallel matrix operations on a shared-memory multiprocessor computer are presented in Table 25. The Ritz vectors φ_i^* ($i = 1, 2, \ldots, n$), resulting at the end of step 1 of Table 24, satisfy the conditions

$$\varphi_i^{*T}\mathbf{M}\varphi_i^* = 1, \quad \varphi_j^{*T}\mathbf{M}\varphi_i^* = 0 \quad (j \neq i)$$

In order to satisfy the orthogonality conditions with respect to the stiffness matrix (i.e. $\varphi_j^{*T}\mathbf{K}\varphi_i^* = 0$ ($j \neq i$)), a tridiagonal matrix $\mathbf{L}_n$ is constructed in step 2 of Table 24. Next, we solve the reduced eigenproblem: $\mathbf{L}_n\mathbf{Q} = \mathbf{Q}\boldsymbol{\Omega}$. The concurrent algorithm developed by Lo & Phillipe [1986] and Farhat & Wilson [1987] for solving the symmetric tridiagonal eigenproblem is adopted in this book. The algorithm is shown in Table 26. The basic idea is to subdivide the eigenvalue spectrum into equal intervals, each assigned to one thread. Each thread computes concurrently the eigenvalues of the reduced system within the interval designated to it. Eigenvectors of the reduced system are then retrieved concurrently.

Finally, we determine the eigenvalues ω_i ($i = 1, 2, \ldots, n$) and the eigenvectors (final Ritz vectors), φ_i ($i = 1, 2, \ldots, n$) for the original system. The eigenvalues are determined from the relation $\omega_i^2 = 1/\Omega_i$.

TABLE 24
Parallel algorithm for the generation of Ritz vectors for use on shared memory multiprocessors

1. Solve for the Ritz vectors φ_i^* ($i = 1, 2, \ldots, n$).
   (a) If $i = 1$, set $\mathbf{v}_{i-1} = \mathbf{F}$. Otherwise, compute $\mathbf{v}_{i-1} = \mathbf{Y}_{i-1}/\beta_{i-1}$ using the normalization algorithm presented in Table 25.
   (b) Solve $\mathbf{K}\hat{\varphi}_i = \mathbf{v}_{i-1}$, using the algorithms described in Chapter IV, where $\mathbf{v}_{i-1}$ represents the load vector.
   (c) If $i = 1$, go to step (f). Otherwise, evaluate the diagonal term of the tridiagonal matrix $\mathbf{L}_n$, $\alpha_{i-1} = \hat{\varphi}_i \mathbf{v}_{i-1}$, using the inner product algorithm shown in Table 25.
   (d) If $i = n$, STOP. Otherwise, compute the scalar quantities $\gamma_j = \varphi_j^{*T}\mathbf{M}\hat{\varphi}_i = \varphi_j^{*T}\mathbf{T}_i$ ($j = 1, 2, \ldots, i-1$) using the matrix multiplication and inner product algorithms shown in Table 25.
   (e) Orthogonalize with respect to $\mathbf{M}$, $\varphi_i^{**} = \hat{\varphi}_i - \sum_{j=1}^{j=i-1} \gamma_j \varphi_j^*$, using the orthogonalization algorithm given in Table 25.
   (f) If $i = 1$, compute the scalar quantity $\kappa_i = \hat{\varphi}_i^T \mathbf{M}\hat{\varphi}_i = \hat{\varphi}_i^T \mathbf{Y}_i$, using the matrix multiplication and inner product algorithms shown in Table 25. Otherwise, compute the scalar quantity $\kappa_i = \varphi_i^{**T}\mathbf{M}\varphi_i^{**} = \varphi_i^{**T}\mathbf{Y}_i$ using the matrix multiplication and inner product algorithms shown in Table 25.
   (g) Evaluate the off-diagonal term of the tridiagonal matrix $\mathbf{L}_n$, $\beta_i = \sqrt{\kappa_i}$.
   (h) Normalize with respect to $\mathbf{M}$. If $i = 1$, compute $\varphi_i^* = \hat{\varphi}_i/\beta_i$ using the normalization algorithm shown in Table 25. Otherwise, compute $\varphi_i^* = \varphi_i^{**}/\beta_i$ using the normalization algorithm shown in Table 25.
2. Construct the tridiagonal matrix $\mathbf{L}_n$ (order n):

$$\mathbf{L}_n = \begin{bmatrix} \alpha_1 & \beta_2 & & & \\ \beta_2 & \alpha_2 & \beta_3 & & \\ & \beta_3 & \alpha_3 & & \\ & & & \alpha_{n-1} & \beta_n \\ & & & \beta_n & \alpha_n \end{bmatrix}$$

An algorithm similar to the normalization algorithm shown in Table 25 is used to perform this step concurrently. The eigenvectors are evaluated from the relation $\mathbf{\Phi} = \mathbf{\Phi}^*\mathbf{Q}$. This matrix multiplication is performed concurrently as shown in Table 25.

6.4 TEMPORAL SOLUTION

At this point, the coupled dynamic equilibrium equation (44) has already been reduced to n uncoupled generalized SDOF (single-degree-of-freedom) systems expressed in modal coordinates (see eqn

TABLE 25
Parallel matrix operations on a shared memory multiprocessor computer

Multiplication of two matrices
If we need to multiply $\mathbf{AB} = \mathbf{C}$, where the matrices $\mathbf{A}$, $\mathbf{B}$, and $\mathbf{C}$ have the dimensions $ROW \times COLUMN$, $COLUMN \times KOLUMN$, and $ROW \times KOLUMN$, respectively, then the following algorithm performs the parallel matrix multiplication on a shared-memory machine.
For $thread = 1, \ldots, num\_threads$, do concurrently
  1. Determine the row location: $row = thread$.
  2. DO sequentially
      For $kolumn = 1, \ldots, KOLUMN$
        For $column = 1, \ldots, COLUMN$
          $C[row][kolumn] = C[row][kolumn]$
                    $+ A[row][column] \times B[column][kolumn]$
        Next *column*.
      Next *kolumn*.
      $row = row + num\_threads$
    WHILE ($row \leq ROW$)

Inner product of two vectors
If we need to multiply $\mathbf{A}^T\mathbf{B} = C$, where both vectors $\mathbf{A}$ and $\mathbf{B}$ have the dimensions $ROW \times 1$, the superscript T refers to the transpose of the vector, and the outcome C is a scalar quantity, then the following algorithm performs the parallel inner product on a shared-memory machine. The vector $\mathbf{D}$ with dimensions $num\_threads \times 1$ is used for temporary storage.
For $thread = 1, \ldots, num\_threads$, do concurrently
  1. Determine the row location: $row = thread$.
  2. DO sequentially
      $D[thread] = D[thread] + A[row] \times B[row]$
      $row = row + num\_threads$
    WHILE ($row \leq ROW$)
  3. For $thread = 1, \ldots, num\_threads$, do sequentially
      $C = C + D[thread]$

Orthogonalization
Consider the vector $\mathbf{C}$ with dimensions $ROW \times 1$. The following algorithm orthogonalizes $\mathbf{C}$ with respect to the columns of a matrix $\mathbf{A}$ with dimensions $ROW \times COLUMN$. The vector $\mathbf{B}$ with dimensions $COLUMN \times 1$ stores the orthogonalization coefficients.
For $thread = 1, \ldots, num\_threads$, do concurrently
  1. Determine the row location: $row = thread$.
  2. DO sequentially
      For $column = 1, \ldots, COLUMN$
        $C[row] = C[row] - A[row][column] \times B[column]$
      Next *column*.
      $row = row + num\_threads$
    WHILE ($row \leq ROW$)

TABLE 25—contd.

Normalization
If we need to normalize a vector **A** with dimensions $ROW \times 1$ with respect to a scalar quantity C then the following algorithm performs the parallel normalization on a shared-memory machine.
For $thread = 1, \ldots, num\_threads$, do concurrently
1. Determine the row location: $row = thread$.
2. DO sequentially
    $A[row] = A[row]/C$
    $row = row + num\_threads$
    WHILE ($row \leq ROW$)

---

(47)). In this work, the Wilson-Θ direct integration method is used for solving eqn (47). The number of eigenmodes n is mapped evenly onto the threads. First, each thread evaluates the load components associated with its assigned modes and shown on the right-hand side of eqn (47). This step involves only pure vector manipulations, which can be evaluated as presented in Table 25. Each thread evaluates the modal displacements for its assigned modes at all the time increments. The algorithm for parallel temporal solution is shown in Table 27.

TABLE 26
Concurrent solution of the reduced eigenvalue problem $\mathbf{L}_n \mathbf{Q} = \mathbf{Q}\mathbf{\Omega}$

For $thread = 1, \ldots, 2$, do concurrently
1. One processor computes Ω_1 and the first column of the modal shape matrix $\mathbf{Q}$ using the inverse iteration method.
2. Another processor concurrently computes Ω_n and the last column of the modal shape matrix $\mathbf{Q}$ using the forward iteration method.

For $thread = 1, \ldots, num\_threads$, do concurrently
1. Define a local interval of search for eigenvalues $[a_{thread}, b_{thread}]$

$$\delta = \frac{\Omega_n - \Omega_1}{num\_threads}, \quad a_{thread} = \Omega_1 + [\delta \times (thread - 1)], \quad b_{thread} = a_{thread} + \delta$$

2. Determine the number of eigenvalues lying in $[a_{thread}, b_{thread}]$ using a Sturm sequence property.
3. Extract all the eigenvalues in $[a_{thread}, b_{thread}]$ using a simple bisection method in conjunction with Sturm sequence property.
4. Compute the corresponding eigenvectors using the inverse shifted iteration method.

TABLE 27
Parallel temporal solution

For *thread* = 1, ..., *num_threads*, do concurrently
1. Determine the mode: *mode* = *thread*.
2. Determine the modal load as defined on the right-hand side of eqn (47). This step involves matrix operations that can be performed concurrently as shown in Table 25.
3. DO sequentially
   For $t = 1, \ldots, num\_time\_increments$
   Evaluate the modal response for the current mode at the current time increment
   Next *t*.
   mode = *mode* + *num_modes*
   WHILE (*mode* ≤ *num_modes*)

6.5 DYNAMIC RESPONSE

The dynamic displacement vectors can be calculated using eqn (45). This step involves matrix multiplication, which can be performed concurrently as presented in Table 25. Each column of the matrix **u** represents the vector of dynamic displacements at a certain time increment. Next, the element forces and stresses corresponding to each vector of dynamic displacements are evaluated. The balance of elements achieved at the end of the initial partitioning stage is used in this step. Each thread evaluates the stiffness matrices of its assigned elements and then evaluates the elements forces and stresses concurrently. This step is outlined in Table 22.

Once the dynamic displacements and stresses at different time increments have been calculated, and the static displacements and stresses evaluated using the algorithms described in Chapter IV, the combined displacements and stresses are computed using superposition. The parallel algorithms described in Chapter V are then used to evaluate the new design vector and consequently the new weight of the structure for this iteration. This iterative procedure continues until no further reduction in the weight of the structure can be achieved. The final weight represents the minimum weight design for the structure.

6.6 APPLICATIONS

The code is applied to the 798-element frame shown in Fig. 73, the 200-bar plane truss shown in Fig. 45, the 90-element braced frame shown in Fig. 39, and the geodesic dome space structure shown in Fig. 42, respectively. The pre-conditions mentioned earlier for the minimum number of elements and nodes within each subdomain are also used in this chapter.

In all examples, the case of the most active displacement constraint is first considered; then the effect of including the set of most active constraints is investigated. In order to assess the convergence properties and stability of the algorithms, the program is allowed to run for 50 iterations for all cases considered in this work. The iteration history for the first 20 iterations is shown below.

A combination of static and dynamic loadings is considered. The static loads for each example are described separately in Section 6.6.1. A horizontal ground acceleration function of $0 \cdot 2g \sin 5 \cdot 236t$ ($0 < t < 1 \cdot 2$ s) is assumed as the dynamic load for all the examples. For all the cases studied here 22 eigenmodes are included in the dynamic analysis.

The optimal designs for the four examples are presented, followed by a study of concurrent performance issues such as speed-up, workload balance, and efficiency. The definitions for speed-up, workload balance, and efficiency are similar to those mentioned in Chapters IV and V.

6.6.1 Optimal design

First, the 798-element frame shown in Fig. 73 is introduced as an example to illustrate the capability of the parallel algorithms in handling large structures. The girders and columns are assumed to have constant lengths of 20 and 12 ft, respectively. The modulus of elasticity is 29 000 ksi, while the specific weight is $0 \cdot 283 \text{ lb/in}^3$. The allowable stresses in the elements are ± 22 ksi. Minimum cross-sectional area is $5 \cdot 0 \text{ in}^2$. Displacements of the nodes are restricted to $\pm 0 \cdot 005h$ in the horizontal direction, where h is the elevation of the node from the ground level. The girders in each story are represented by one variable and the columns by another variable. This reduces the number of design variables to 120. The girders are subjected to a uniform distributed load of 120 lb/in acting downwards along their spans. For the moment-resisting elements, the moments of inertia are taken to be 75 times the cross-sectional areas, and the section moduli as 9 times the cross-sectional areas. These relationships appear to be representative of commonly used wide flange shapes, and are chosen in order to maintain linearity between element stiffness matrices and

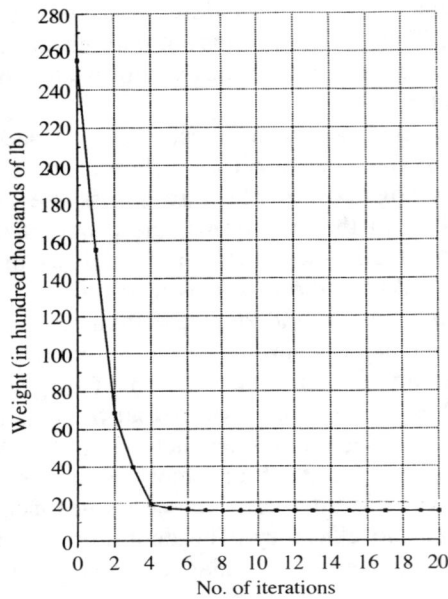

Fig. 92. Iteration history for the 798-element frame.

element cross-sectional areas. Figure 92 shows the iteration history when only the most active displacement constraint is considered. An optimal design of 1537 kips is achieved. Including the set of most active displacement constraints does not change the results.

For the 200-bar plane truss shown in Fig. 45, all members are made of steel with $E = 30 \times 10^6$ psi and $\rho = 0 \cdot 283$ lb/in^3. The maximum allowable stresses are ± 29 ksi. Minimum cross-sectional areas of $1 \cdot 0$ in^2 are allowed. The structure is symmetric about the vertical centerline, thus reducing the number of design variables to 105. The truss is subjected to a set of vertical loads with values of 10 kips acting downwards at the nodes along the vertical lines 1–71, 2–76, 3–73, 4–77, and 5–75. Displacement limits of $0 \cdot 005h$ in are imposed on all nodes in the horizontal direction, where h is the elevation of the node above the ground. Figure 93 shows the iteration history. An optimal design of 22 156 lb is achieved. When the set of most active displacement constraints is considered, no improvement in the optimal weight is achieved.

For the 90-element braced frame shown in Fig. 39, the modulus of

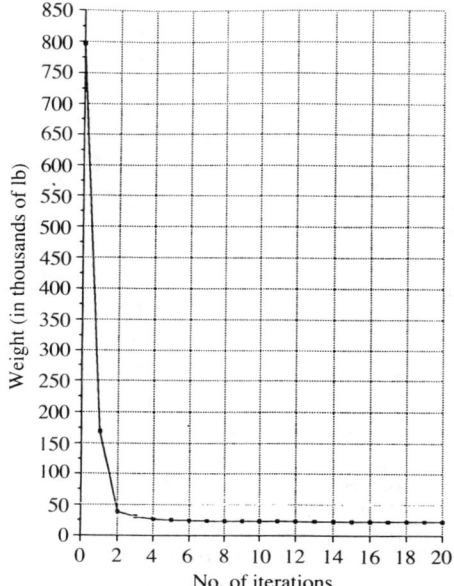

Fig. 93. Iteration history for the 200-bar plane truss.

elasticity of the frame material is 29 000 ksi and the specific weight is 0·283 lb/in^3. The allowable stresses in the moment-resisting elements are ±29 ksi, while those in the truss elements (bracings) are ±20 ksi. The minimum cross-sectional area is 0·5 in^2. Displacements of the nodes are restricted to ±0·005h in the horizontal direction, where h is the elevation of the node above the ground. The structure is symmetric about the vertical centerline. This reduces the number of design variables to 45. The horizontal moment-resisting elements are subjected to a uniform distributed load of 180 lb/in acting downwards along their spans. The same relationships relating the moments of inertia and the section moduli to the cross-sectional areas described for the 798-element frame are used in this example. Figure 94 shows the iteration history when only the most active displacement constraint is considered. An optimal design of 78 907 lb is achieved. When the set of most active displacement constraints is considered, no improvement in the optimal minimum weight is achieved.

Finally, the optimal design of the geodesic dome space truss shown in Fig. 42 is presented. The modulus of elasticity is 10 000 ksi and the

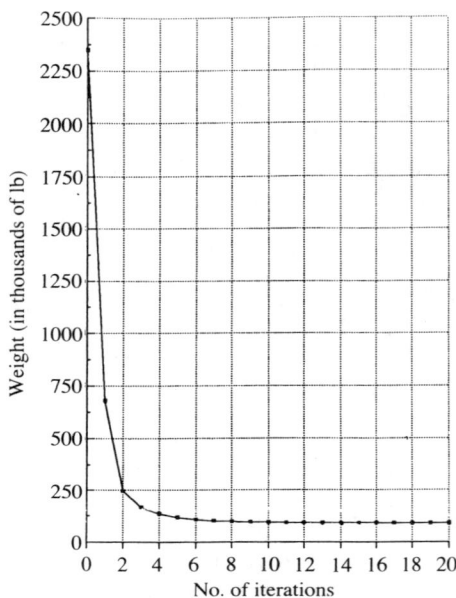

Fig. 94. Iteration history for the 90-element braced frame.

specific weight of the truss material is $0 \cdot 1$ lb/in^3. Allowable stresses are limited to ± 25 ksi. A minimum cross-sectional area of $0 \cdot 1$ in^2 is allowed. Displacement limits of $\pm 0 \cdot 1$ in are imposed on all nodes in the x, y, and z directions. The structure is symmetric about the line connecting nodes 5 and 57 and that connecting nodes 12 and 50. This reduces the number of design variables to 36. A load of 1 kip (1000 lb) is acting downwards in the z direction at all the nodes. An optimal design of $200 \cdot 4$ lb is obtained when only the most active displacement constraint is considered. When the set of most active displacement constraints is considered, a minimum weight of $184 \cdot 3$ lb is achieved. Figure 95 shows the iteration history for this case.

At this point, we note that the case when only the most active displacement constraint is included in the formulation yields the optimal design for most problems. Even for the case of the geodesic dome space truss, a 'near' optimum was obtained. Therefore, we shall consider this case to be representative of the algorithm's performance. Another characteristic of this case is that it results in iterations with almost identical amount of workloads. Consequently, the performance

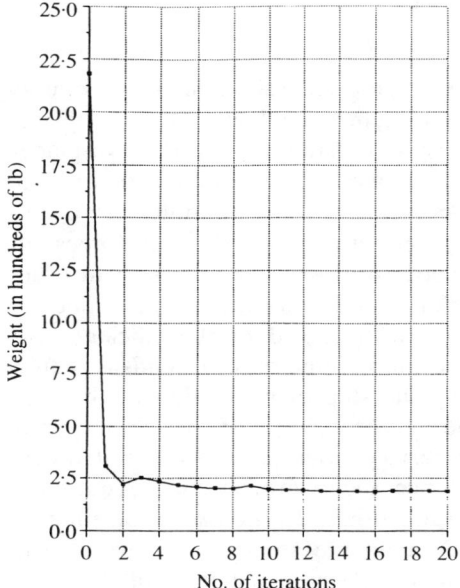

Fig. 95. Iteration history for the geodesic dome space truss.

of the algorithm can be assessed through the study of a single iteration. A single optimization iteration consists of two phases: analysis and redesign. In the following, the concurrent performance of the computational steps in a single iteration for the 798-element frame is discussed. The overall performance within an iteration is also presented for the four examples described earlier.

Note that the computational time consumed in executing the sequential code tends to be dominated by the step of evaluation of the Ritz vectors, because in this step the entire set of linear equations corresponding to the nodal degrees of freedom needs to be solved a number of times equal to the number of eigenmodes. The time history approach for solving the dynamic problem can also consume a significant execution time. As the number of time increments involved increases, the participation of the steps of the temporal solution, dynamic response, maximum displacements, and maximum stresses in the execution time increases.

6.6.2 Static analysis

The seven steps for static analysis, namely, assembling the structure stiffness matrix, setting up the load vector, applying the boundary conditions, condensation of the non-interface displacement degrees of freedom onto the interface ones, solution of the interface displacement degrees of freedom, retrieving the non-interface displacement degrees of freedom, and computation of element forces and stresses, are visited during the course of analysing the structure under static loads. The step of solution of the interface linear equations represents the bottleneck situation with regard to the efficiency of the concurrent performance. For example, one set of threads is required for achieving concurrency during the step of evaluation of the element forces and stresses, while the number of sets of threads required for the step of solution of the interface linear equations is 9 times the *number of interface nodes*. Since the overhead time required for creating the threads is added sequentially to the execution time of the concurrent application, its effect on the speed-up is severe. This is particularly true for small structures, where the ratio of the number of interface nodes to the number of internal nodes is relatively large. Figure 96 shows how the number of interface nodes increases with increasing number of processors in use for the 798-element plane truss. Figure 97 shows the speed-ups for the cases of the 'threads' and the 'workload balance' (i.e. when the overhead time required for creating the threads is neglected) for the 798-element frame. Figure 98 shows the speed-ups for the cases of the threads and the workload balance for the 200-bar plane truss. The effect of the bottleneck situation on reducing the speed-up is particularly evident for small structures.

6.6.3 Assembling the structure mass matrix

The speed-ups for assembling the structure consistent mass matrix for the 798-element frame are shown in Fig. 99. The greater the ratio of the time required for creating the threads to accomplish a certain step to the sequential time required for executing the step, the more there is an offset from the theoretical linear speed-up case. When the overhead required for creating the threads is neglected, the speed-up curve is practically linear (the workload balance curve).

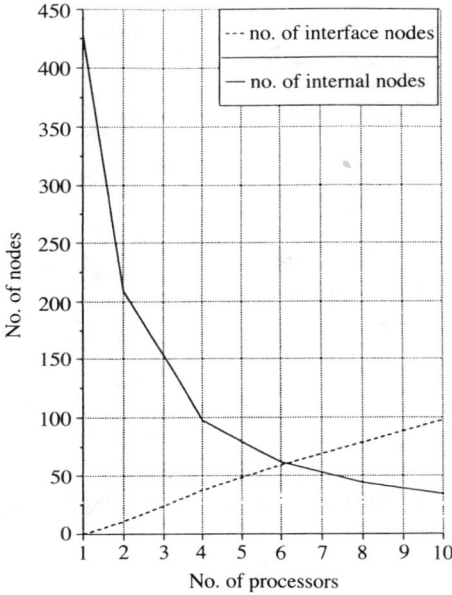

Fig. 96. Numbers of interface and internal nodes for the 798-element frame.

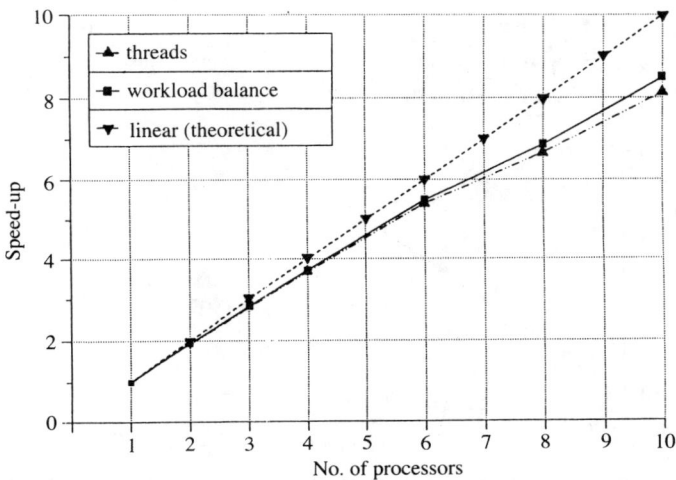

Fig. 97. Speed-ups for the static analysis for the 798-element frame.

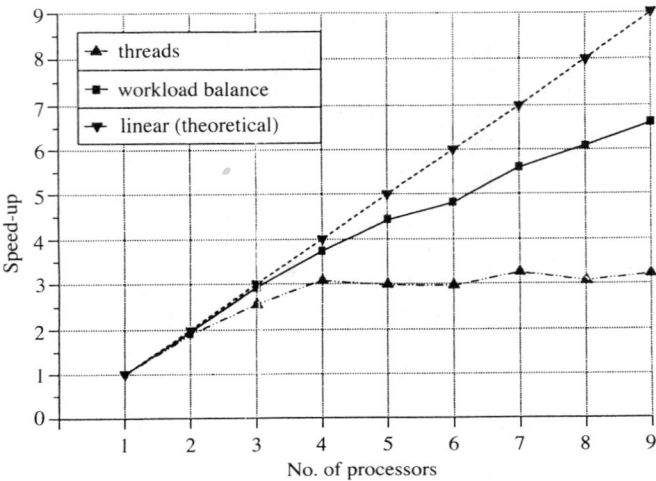

Fig. 98. Speed-ups for the static analysis for the 200-bar plane truss.

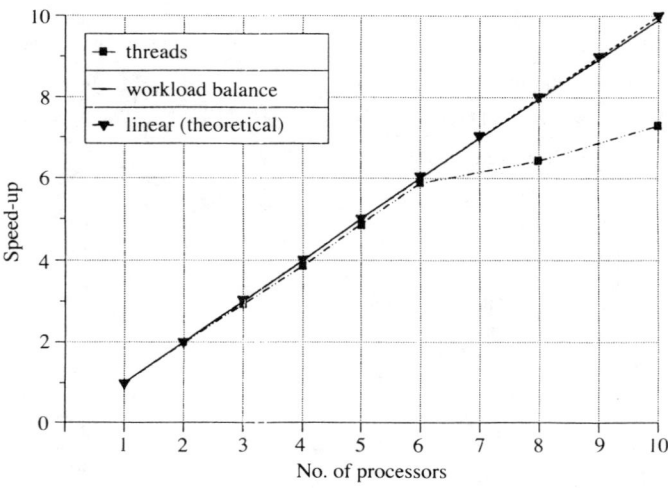

Fig. 99. Speed-ups for assembling of the structure mass matrix for the 798-element frame.

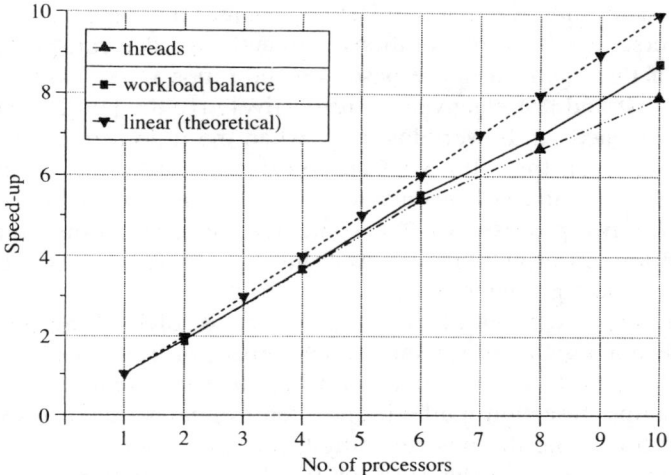

Fig. 100. Speed-ups for evaluation of the Ritz vectors for the 798-element frame.

6.6.4. Evaluation of the Ritz vectors

In this step, a number of Ritz vectors equal the number of eigenmodes and orthonormal with respect to the mass matrix is generated. Each vector requires the solution of the linear system of equations for the degrees of freedom of the structure in addition to several matrix operations such as scaling and inner products. Solution of the linear system of equations a number of times equal to the number of eigenmodes reduces the speed-up due to the large number of threads created in this case. This effect is more significant for small structures. Figure 100 shows the speed-up curves for the 'threads' and 'workload balance' cases for the 798-element frame. Figure 101 shows the results for the 200-bar plane truss.

6.6.5 Solution of the reduced eigenproblem

In order to orthogonalize the Ritz vectors with respect to the structure stiffness matrix, a reduced eigenproblem $L_n Q = Q\Omega$ is formulated from the coefficients generated during the course of the previous step of evaluating the Ritz vectors. The reduced eigenproblem is solved using the algorithm described in Table 26. First, two processors are used to determine the first and last eigenvalues and eigenvectors using the inverse and forward iterations, respectively. The rates of convergence of the eigenvalues are based on the ratios Ω_1/Ω_2 and Ω_{n-1}/Ω_n for the first and last eigenvalues, respectively (Bathe [1982]). The rate of convergence can be very low (e.g. when the ratio is 0·9999) or very high (e.g. when the ratio is 0·01), and this is not known in advance. When the first and last eigenvalues have different convergence rates, one of the two processors will become idle sooner than the other, and the multiprocessor will function as a single processor until convergence of the second eigenvalue occurs.

Next, the spectrum of the eigenvalues is divided as evenly as possible among the processors. Each processor retrieves the eigenvalues lying in its assigned interval using the Sturm sequence property and a simple bisection method, and then evaluates the corresponding eigenvectors using the inverse shifted iteration. Figure 102 shows the speed-ups achieved in this step for the 798-element frame. The poor speed-ups indicate the fact that only one processor will be running during the convergence to the first or last eigenvalue. In addition, the fact that the eigenvalues are not equally balanced among the intervals can contribute to the offset of the curve. However, this step consumes

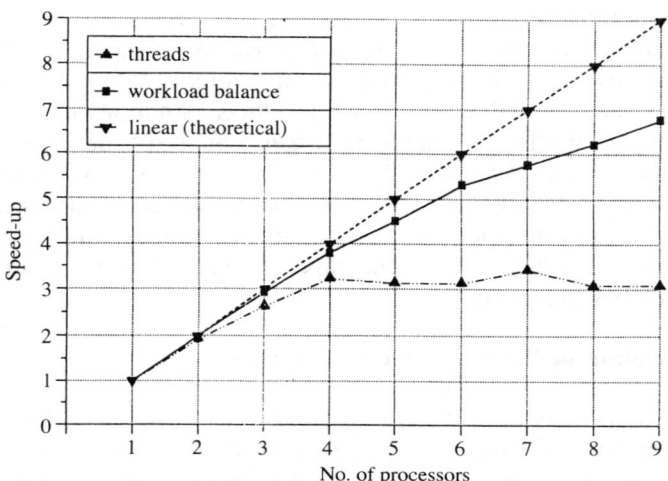

Fig. 101. Speed-ups for evaluation of the Ritz vectors for the 200-bar plane truss.

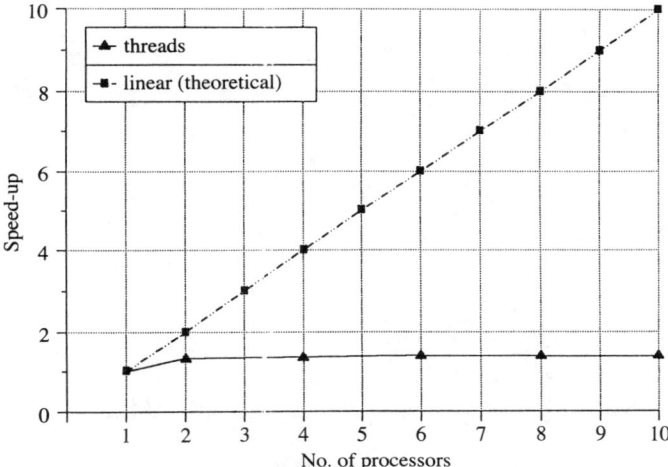

Fig. 102. Speed-ups for solution of the reduced eigenproblem for the 798-element frame.

about 3% or less of the total execution time for the examples considered in this book, and therefore its effect on the overall performance is not significant.

6.6.6 Computation of the eigenvectors (final Ritz vectors)

The eigenvectors of the original system (or the final Ritz vectors) satisfying the conditions of orthonormality with respect to the structure mass matrix and orthogonality with respect to the structure stiffness matrix are determined in this step, which involves multiplication of the matrix of the Ritz vectors, Φ^*, by the matrix of eigenvectors of the reduced eigenproblem, Q. Only one set of threads is required to accomplish this step. Figure 103 shows the speed-ups attained for the 798-element frame for the case of threads.

6.6.7 Temporal solution

The equation $\ddot{z}_i + 2c_i\omega_i\dot{z}_i + \omega_i^2 z_i = D_i^*$ for the modal displacements is solved in this step. The Wilson-Θ method (Bathe [1982]) is used in this work for finding the values of z. Only one set of threads is

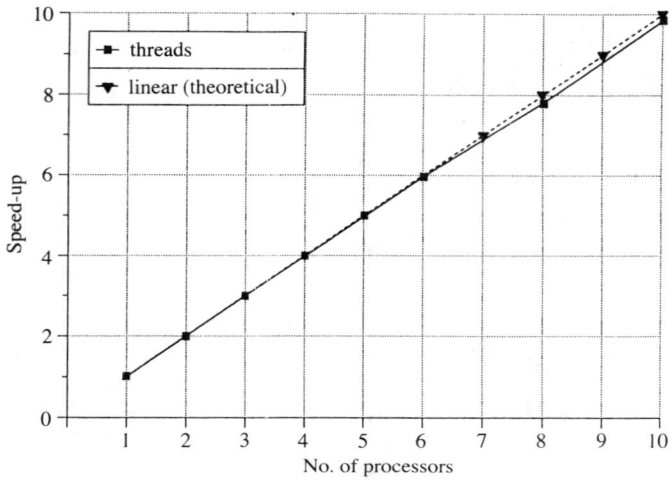

Fig. 103. Speed-ups for the computation of the eigenvectors (final Ritz vectors) for the 798-element frame.

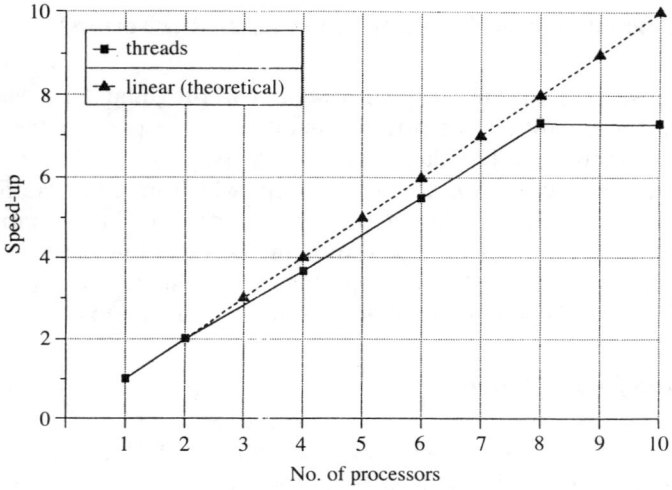

Fig. 104. Speed-ups for the temporal solution step for the 798-element frame.

required. The number of modes is mapped as evenly as possible to the threads. For each one of its assigned eigenmodes, the thread evaluates the right-hand side of the equation using the matrix multiplication algorithm shown in Table 25 and then calculates the modal response at the given time increments.

Figure 104 shows the speed-ups achieved for the temporal solution step for the 798-element frame. The 'workload balance' curve is practially the same as the 'threads' curve, and is therefore not shown in this figure. Figure 104 also shows that no improvement is achieved in speed-up for the cases of 9 and 10 processors compared with the case when 8 processors are in use. This is due to the fact that the number of eigenmodes assigned to the processors for the cases of 9 and 10 processors is not less than the number assigned to them for the case of 8 processors (some processors are assigned 3 eigenmodes while the rest are assigned 2 eigenmodes). Note that the time consumed in this step, and consequently its contribution to the overall execution time, increases as the number of time increments considered for the eigenmodes increases.

6.6.8 Dynamic response

The dynamic displacements, forces, and stresses are evaluated in this step. The computation of the dynamic displacements involves matrix multiplication. One set of threads is required to perform the multiplication concurrently as shown in Table 25. Another set of threads is required for finding the element forces and stresses corresponding to the dynamic displacement vectors. The speed-up achieved in this step for the case of threads for the 798-element frame is practically the same as the case shown in Fig. 103. Note that the contribution of the time consumed in this step to the overall execution time increases as the number of time increments considered for evaluating the dynamic displacements and stresses increases.

6.6.9 Maximum response

The maximum displacements and stresses due to the static and dynamic loadings are computed in this step. First, a set of threads is created for finding the maximum displacement ratio. The balance of nodes is used during this process. Next, another set of threads is used to locate the most active displacement constraints. The balance of

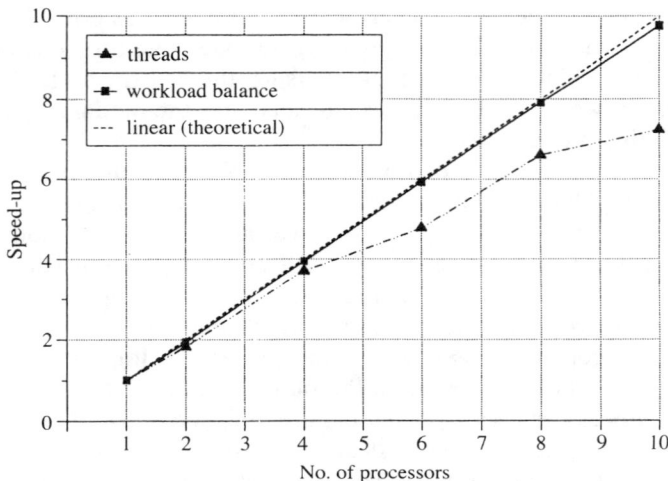

Fig. 105. Speed-ups for evaluation of the maximum response for the 798-element frame.

nodes is also used in this case. The maximum stress ratio is then determined. Two sets of threads are required for finding this ratio. The balance of elements achieved at the end of the initial partitioning stage is used with the first set, while a balance of design variables is used with the second set. Figure 105 shows the speed-ups achieved in this step for the 798-element. The 'workload balance' curve is practically linear. Note that the contribution of the execution time consumed in this step to the overall execution time increases as the number of time increments considered for evaluating the dynamic displacements and stresses increases.

6.6.10 Redesign

After the maximum response has been evaluated, the structure is analysed a number of times equal to the number of active displacement constraints; each time, the loading case is a unit load acting in the direction of the corresponding active displacement constraint. Next, the gradients of the active displacement constraints with respect to the design variables are computed. Finally, the new vector of design variables is calculated. Figure 106 shows the speed-ups achieved in this step for the 798-element frame. The effect of the solution of the

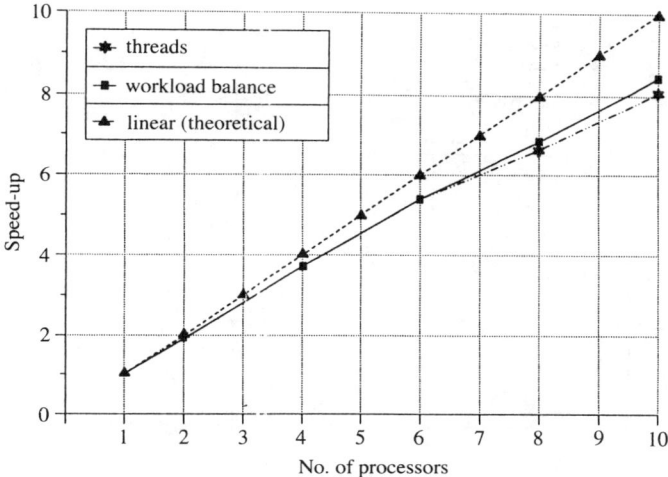

Fig. 106. Speed-ups for evaluation of the gradients for the 798-element frame.

interface linear equations on reducing the speed-ups becomes more noticeable when the size of the structure is small. Figure 107 shows the speed-ups achieved in this step for the 200-bar plane truss.

6.6.11 Overall speed-up, workload balance, and efficiency

The overall performance of the parallel algorithms for optimization of structures subjected to combined static and dynamic loadings is now assessed. A comparison of the overall speed-ups for a single iteration for the 798-plane frame, the 200-bar plane truss, the 90-element braced frame, and the geodesic dome space truss for the case of 'threads' is shown in Fig. 108. Execution is automatically terminated when the minimum number of elements (nine) or internal nodes (three) is not satisfied within the subdomains. This represents the case of ten or more subdomains for the 200-bar plane truss, five or more subdomains for the geodesic dome space truss, and eight or more subdomains for the 90-element plane frame.

Table 28 lists the speed-ups for the 'threads' case for the four examples considered for the cases of 4, 7, 9, and 10 processors. The number of degrees of freedom for each structure is also included in this table. Figure 108 and Table 28 show that better speed-ups are

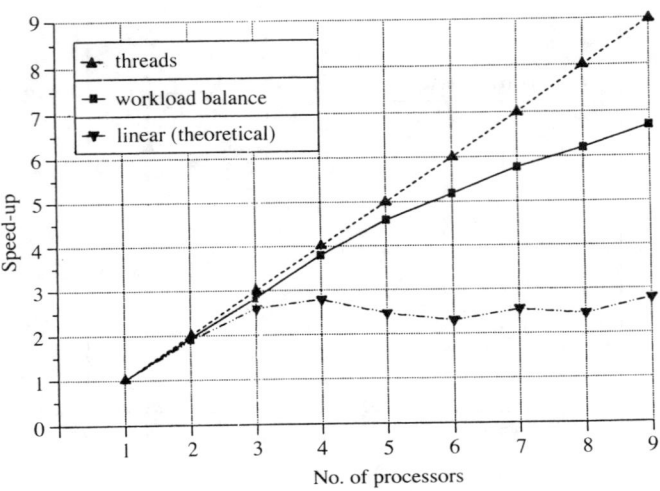

Fig. 107. Speed-ups for evaluation of the gradients for the 200-bar plane truss.

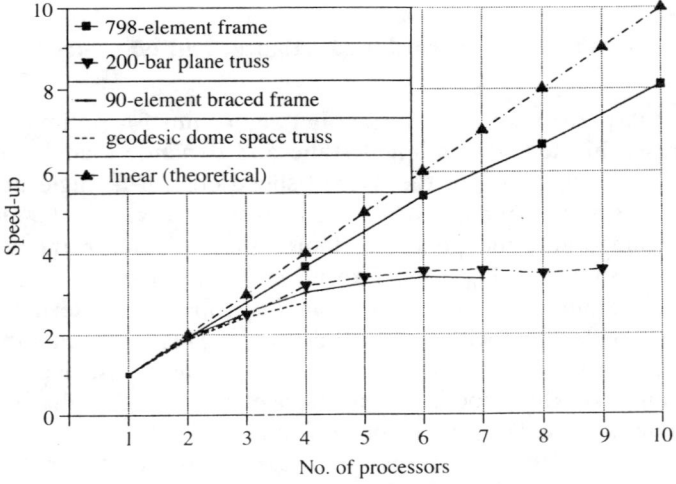

Fig. 108. Overall speed-ups for the case of 'threads' for the 798-element frame, the 200-bar plane truss, the 90-element braced frame, and the geodesic dome space truss.

TABLE 28
Overall speed-up for the example problems (sinusoidal dynamic loading)

Example	Number of degrees of freedom	Number of processors			
		4	7	9	10
798-element frame	1287	3·67	6·03	7·39	8·11
200-bar plane truss	150	3·20	3·60	3·60	—
90-element braced frame	135	3·03	3·37	—	—
Geodesic dome space truss	111	2·79	—	—	—

achieved for larger structures when the same number of processors is used. On one hand, this is due to the fact that the overhead time required for creating a thread tends to be less significant compared with the time required for the thread to execute its task as the size of the task assigned to each thread increases. On the other hand, the percentage of the time consumed in the fine-grained parallelism step (solution of the linear interface equations) tends to be less significant compared with the overall execution time as the size of the problem increases. This explains why, for example, the speed-up curve for the 798-element frame is higher than that for the 200-bar plane in Fig. 108. When the overhead time required for creating the threads and the time required for solving the set of linear interface equations is large relative to the total execution time, the concurrent application fails to pick up additional speed-ups as the number of processors is increased. This is observed in Fig. 108 for the 200-bar plane truss for the cases of more than seven processors and for the 90-element plane frame for the case of more than six processors.

Figure 109 shows the overall workload balance for a single iteration of the optimization process for the 798-plane frame, the 200-bar plane truss, the 90-element braced frame, and the geodesic dome space truss. For 1–4 processors, all curves practically coincide. The offset of the workload balance curves in Fig. 109 from the theoretical linear speed-up case is mainly due to imbalance of workload. The effect of this imbalance becomes more significant as the number of threads (and processors) increases and consequently the task of each thread becomes smaller. Finally, the overall efficiency for the four examples considered here is shown in Fig. 110. An overall efficiency of 85% was achieved for the 798-element frame for the case of 10 processors.

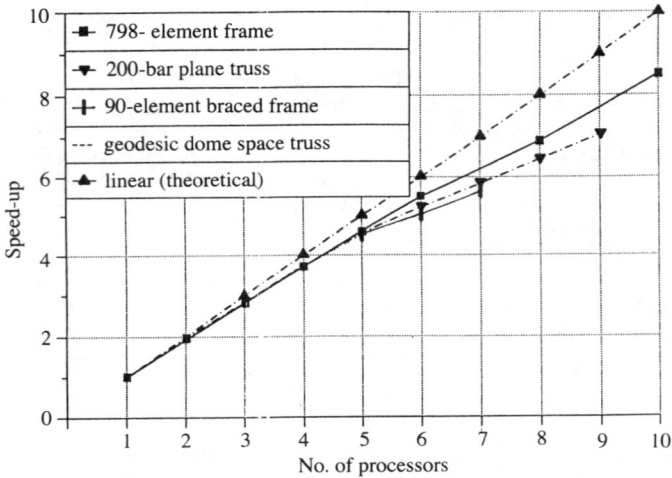

Fig. 109. Overall speed-ups for the case of 'workload balance' for the 798-element frame, the 200-bar plane truss, the 90-element braced frame, and the geodesic dome space truss.

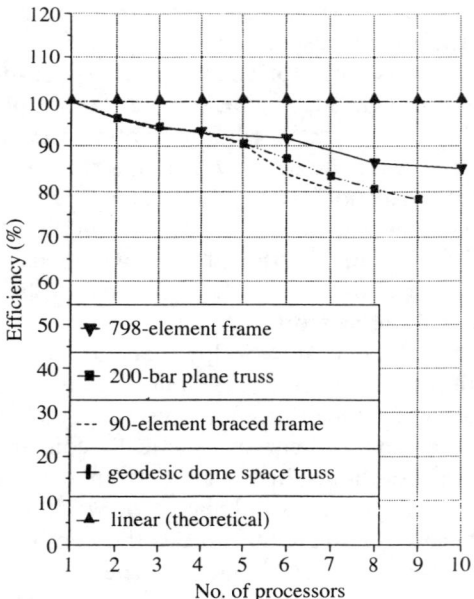

Fig. 110. Overall efficiencies for the 798-element frame, the 200-bar plane truss, the 90-element braced frame, and the geodesic dome space truss.

6.6.12 El Centro earthquake

In order to study the effect of the time history analysis on the speed-up, the optimal design of structures subjected to the El Centro earthquake (California, 1940) is considered. The ground accelerations recorded during this earthquake are shown in Fig. 111. Compared with the sinusoidal dynamic loading, more time is spent on the steps of the temporal solution, dynamic response, and maximum response. Better speed-ups are expected, since these steps are parallelized effectively. Figure 112 shows the improvement achieved in the speed-up for the 'threads' case when the El Centro record is used compared with the sinusoidal dynamic loading, for the 200-bar plane truss. However, less improvement is achieved for the case of the 798-element frame, since the ratio of the time spent on the aforementioned three time history steps to the overall execution time is smaller for this larger structure (Fig. 113). Table 29 lists the speed-ups achieved for the case of 'threads' for the four example problems considered in this work for the El Centro earthquake for the cases of 4, 7, 9, and 10 processors.

6.7 SUMMARY AND CONCLUSIONS

Parallel algorithms and stratagems have been presented for optimization of framed structures subjected to combined static and dynamic

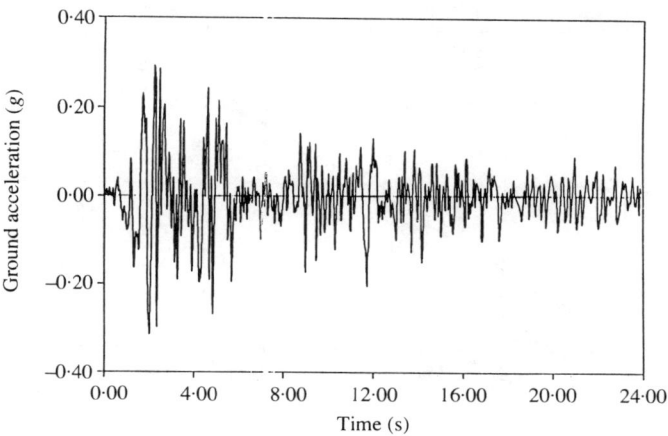

Fig. 111. Ground accelerations recorded during the El Centro earthquake (California, 1940).

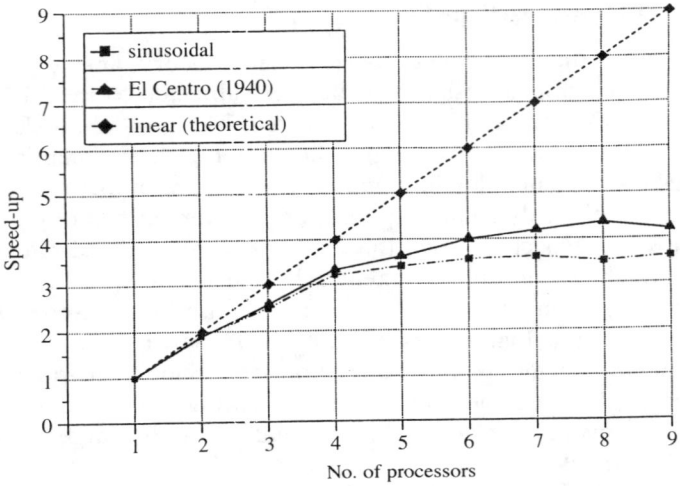

Fig. 112. Overall speed-ups for the case of 'threads' for the 200-bar plane truss.

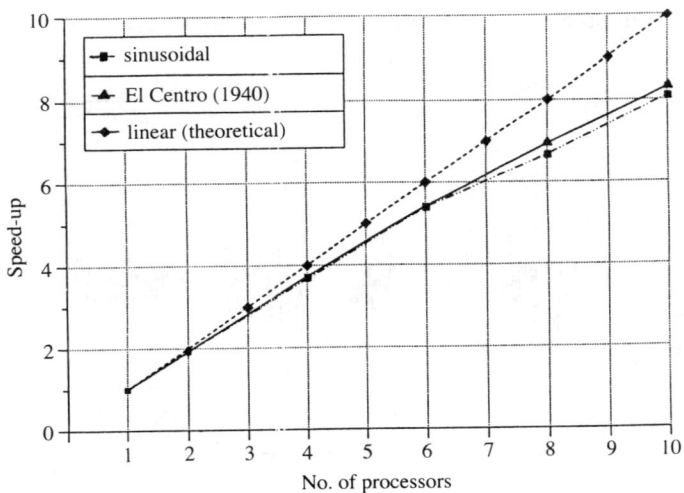

Fig. 113. Overall speed-ups for the case of 'threads' for the 798-element plane frame.

TABLE 29
Overall speed-up for the example problems (El Centro earthquake)

Example	Number of degrees of freedom	Number of processors			
		4	7	9	10
798-element frame	1287	3·73	6·17	7·62	8·31
200-bar plane truss	150	3·32	4·18	4·21	—
90-element braced frame	135	3·25	3·86	—	—
Geodesic dome space truss	111	3·02	—	—	—

loadings. Parallelism is achieved through the notion of cheap concurrency and the concept of threads. Emphasis is directed towards attaining a workload balance during each computational step of the solution process. The algorithms have been implemented in C on an Encore Multimax shared-memory computer. The step of solving the interface linear equations represents the bottleneck for the overall performance owing to the need for the creation of a large number of threads. This shows the merit of the overall approach presented in this book in reducing the number of linear equations required to be solved simultaneously to those only associated with the interface through the concept of substructuring. For large structures, however, the effect of this bottleneck situation is less severe, and better performance is achieved, since the percentage of the sequential time required for solving the interface linear equations is small compared with the overall sequential time. In addition, the time history analysis can play a positive role in improving the efficiency of the concurrent application. The more time increments that are required during the dynamic analysis, the greater are the contributions of the steps of the temporal solution, dynamic response, and maximum response to the overall execution time. Consequently, better speed-ups are achieved, because these steps are parallelized effectively without the need for creating a large number of threads.

The optimization algorithms are built mostly around the steps involved in the analysis phase. In concurrent processing terms, this makes the concurrent optimization algorithms perform as well as the concurrent analysis algorithms. In optimization terms, the efficiency of the algorithms is independent of the number of design variables and behavior constraints to a large extent, thus allowing problems with a large number of design variables and behavior constraints to be handled effectively. As clearly seen in Fig. 108, the concurrent algorithms presented in this book are particularly effective for optimization of large structures.

Chapter VII

Conclusions

7.1 SUMMARY AND CONCLUSIONS

1. Parallel algorithms have been developed for the analysis and optimization of structures under static and dynamic loadings. The parallel algorithms are implemented in the C programming language on an Encore Multimax shared-memory multiprocessor computer. Concurrency is achieved through the use of the notion of cheap concurrency and the concept of threads.

2. The following conclusions are drawn based on experimentation with the Encore Parallel Threads (EPT) on an Encore Multimax with 12 processors.

(a) The amount and frequency of shared memory accesses such as reads and writes have an insignificant effect on the concurrent performance of the algorithms. In other words, no communication overhead is involved.

(b) The overhead time required for synchronization slows down the speed-up expected from the Encore Multimax. This is mainly due to the overhead time required for the invocation of semaphores.

(c) The overhead time required for the creation of threads is added sequentially to the execution time. The effect of this overhead on reducing the speed-up is more significant when the overhead time required to create a thread is large compared with the execution time of the task to which the thread is assigned.

(d) Having a number of threads equal to or less than the number of processors in use reduces the effect of the overhead time due to context switching.

3. A three-stage substructuring algorithm has been developed with the following features:

(a) It provides maximum concurrency during major portions of the computations without excessive interprocessor communications or excessive creation of threads. The only exception is in the step of solution of the interface linear equations.

(b) It uses different workload balancing stratagems, thus providing different steps with their most appropriate workload balancing schemes.

(c) It minimizes the number of interface nodes where an excessive number of threads is required to solve the set of interface linear equations.

4. The main conclusions drawn from the algorithms for concurrent analysis of structures are as follows (Adeli & Kamal [1990a, 1992a, b]):

(a) Synchronization using semaphores is less efficient than using extra storage for the locations of the shared memory subjected to a racing condition followed by an updating step.

(b) Solution of the interface linear equations represents the bottleneck situation for the entire solution process. This is due to the fact that a large number of threads is required in this step.

(c) The efficient storage schemes introduced in this book allow for large structures to be handled effectively.

5. The main features of the optimization formulation developed in this work and its concurrent algorithms are as follows (Adeli & Kamal [1990b, c, 1992c, d]):

(a) It is built mostly around the steps involved in the structural analysis.

(b) One active displacement constraint is generally sufficient for achieving the optimum design. This means small storage requirements and fewer computations.

(c) The formulation is independent of the number of design variables and behavior constraints to a large extent, thus allowing structures with large numbers of such optimization parameters to be handled effectively.

6. The following conclusions can be drawn with regard to the

dynamic analysis and its concurrent algorithms (Adeli & Kamal [1992e, f]):

(a) The time spent on the evaluation of the Ritz vectors tends to dominate the execution time. This is due to the large number of times the set of linear equations is solved in this step.
(b) The time history analysis can play a positive role in improving the efficiency of the concurrent application. The more time increments that are required during the dynamic analysis, the greater are the contributions of the steps of the temporal solution, dynamic response, and maximum response to execution time. Consequently, better speed-ups are expected, since these steps parallelize well without the need for creating a large number of threads

7. The efficiency of the algorithms for structural analysis, optimization under static loading, and optimization under dynamic loading increases with the size of the structure. This is due to the fact that the time consumed in the 'fine-grained' parallelism step of solution of the interface linear equations tends to be less significant compared with the overall execution time as the size of the problem increases. The significant conclusion is that the algorithms developed in this research are particularly efficient for the analysis and optimization of large structures such as high-rise buildings and space stations.

7.2 FUTURE RESEARCH

Additional research can be done on the issue of workload balance achieved by the substructuring algorithm. For example, a study of the effect of the size of the bandwidths of the subdomain matrices on the speed-up can provide additional insight.

We have presented parallel algorithms for the optimization of structures using the optimality criteria approach. Parallel algorithms can be developed for structural optimization using mathematical programming approaches such as the general geometric programming technique (Adeli & Kamal [1986]). Other techniques should also be investigated for solving the eigenvalue problem.

Vector MIMD machines such as the Cray Y-MP 8/864 provide both vector and parallel processing capabilities. The Cray Y-MP 8/864 supercomputer has eight processors. Therefore, one can combine vector processing with parallel processing to achieve maximum processing speed. The simultaneous use of more than one processor is referred to as multitasking. Concurrent processing or multitasking on the Cray Y-MP 8/864 is performed by macrotasking, microtasking, and autotasking. Macrotasking is performed at function level. Normally, major tasks that can be processed concurrently are macrotasked. Macrotasking is implemented by function calls, and is suitable for tasks requiring large processing time because its overhead is large compared with that of microtasking. Macrotasked tasks should be identified when the general concurrent algorithm is developed. Microtasking is parallel processing at the loop level. Autotasking is the automatic distribution of tasks to multiple processors by compiler. The senior author and his associates are currently developing vector-parallel algorithms for the optimization of large structures by judicious combination of vectorization, microtasking, and multitasking (Adeli [1992a, b], Vishnubhotla & Adeli [1992], Adeli et al. [1993]).

Appendix: Sample Code

The following sample C code shows the Encore Parallel Threads function calls required for the creation of threads and synchronization, respectively.

CREATION OF THREADS

```
/* create a # of threads equal to the number of processors */
for (p = 1; p < = np; p + + )
{
    /* assemble structure stiffness matrix and load vector */
    tcb_array [p - 1] = THREADcreate(assemble, p, 0, 0, 200*1024, 2);
}
```

where *np* is the number of processors, *tcb_array* is the thread control block array, and *assemble* is the function for which the threads are created.

SYNCHRONIZATION USING SEMAPHORES

```
THREADpsem(sem[datum + kk - 1]);
s_stiff[ii - 1][kk - 1] + = e_stiff[jj - 1][ll - 1];
THREADvsem(sem[datum + kk - 1]);
```

where *THREADpsem* is the function call that invokes the P operation, *THREADvsem* is the function call that invokes the V operation, *sem* is the array of semaphores, *s_stiff* is the structure stiffness matrix, and *e_stiff* is the element stiffness matrix.

References

Adams, L. M. [1985]. Reordering computations for parallel execution. *Communications in Applied Numerical Methods*, **2**, 263–71.
Adeli, H., ed. [1992a]. *Parallel Processing in Computational Mechanics*, Marcel Dekker, New York.
Adeli, H., ed. [1992b]. *Supercomputing in Engineering Analysis*. Marcel Dekker, New York.
Adeli, H. & Kamal, O. [1986]. Efficient optimization of space trusses. *Computers and Structures*, **24**, 501–11.
Adeli, H. & Kamal, O. [1988]. Parallel algorithms for structural optimization under dynamic loading. In *Parallel and Distributed Processing in Structural Engineering*, ed. H. Adeli. American Society of Civil Engineers, New York, pp. 27–52.
Adeli, H. & Kamal, O. [1989a]. Parallel structural analysis using threads. *Microcomputers in Civil Engineering*, **4**, 133–47.
Adeli, H. & Kamal, O. [1989b]. Structural analysis on a parallel machine. In *Developments in Mechanics*, Vol. 15, ed. J. B. Ligon et al. Michigan Technological University, Houghton, Michigan, pp. 145–46.
Adeli, H. & Kamal, O. [1990a]. Concurrent analysis of structures on a shared memory machine. In *Proceedings of the 10th International Conference on Distributed Computing Systems, Paris, 28 May–1 June 1990*.
Adeli, H. & Kamal, O. [1990b]. Parallel stratagems for optimization of large structural systems. In *Proceedings of the IEEE International Conference on Systems Engineering, Pittsburgh, Pennsylvania, 9–11 August 1990*, pp. 21–4.
Adeli, H. & Kamal, O. [1990c]. Optimization of large structures on multiprocessor machines. In *Proceedings of the 2nd IEEE Workshop on Future Trends of Distributed Computing Systems, Cairo, 30 September–2 October 1990*, pp. 290–5.
Adeli, H. & Kamal, O. [1991]. Parallel algorithms for stress analysis on shared-memory multiprocessors. In *Proceedings of the 1st International Conference of the Austrian Center for Parallel Computation, Salzburg, 30 September–2 October*.

Adeli, H. & Kamal, O. [1992a]. Concurrent analysis of large structures—I—Algorithms. *Computers and Structures,* **42,** 413–24.
Adeli, H. & Kamal, O. [1992b]. Concurrent analysis of large structures—II—Applications. *Computers and Structures,* **42,** 425–32.
Adeli, H. & Kamal, O. [1992c]. Concurrent optimization of large structures: Part I—Algorithms. *Journal of Aerospace Engineering, ASCE,* **5,** 79–90.
Adeli, H. & Kamal, O. [1992d]. Concurrent optimization of large structures: Part II—Applications. *Journal of Aerospace Engineering, ASCE,* **5,** 91–110.
Adeli, H. & Kamal, O. [1992e]. Concurrent optimization of structures under dynamic loading—Part I—Algorithms. *International Journal of Mini and Microcomputers,* **14.**
Adeli, H. & Kamal, O. [1992f]. Concurrent optimization of structures under dynamic loading—Part II—Applications. *International Journal of Mini and Microcomputers,* **14.**
Adeli, H. & Vishnubhotla, P. [1987]. Parallel processing. *Microcomputers in Civil Engineering,* **2,** 257–69.
Adeli, H. & Vishnubhotla, P. [1992]. Parallel processing and parallel machines. In *Parallel Processing in Computational Mechanics,* ed. H. Adeli. Marcel Dekker, New York, pp. 1–20.
Adeli, H., Kamat, M., Kulkarni, !., & Vanluchene, D. [1993]. Review of high performance computing methods in structural mechanics. *Journal of Aerospace Engineering, ASCE,* **5** (in press).
Bathe, K. J. [1982]. *Finite Element Procedures in Engineering Analysis.* Prentice-Hall, Englewood Cliffs, NJ.
Bostic, S. & Fulton, R. [1988]. Vibrational analysis using a parallel Lanczos method. In *Parallel and Distributed Processing in Structural Enginereing,* ed. H. Adeli. American Society of Civil Engineers, New York, pp. 1–12.
Burden, R. L., Faires, J. D. & Reynolds, A. C. [1981]. *Numerical Analysis,* 2nd edn. Prindle, Weber, and Schmidt, Boston, MA.
Cheng, F. Y., Srifuengfung, D. & Sheng, L. H. [1981]. ODSEWS-2D optimal design of static, earthquake, and wind steel structures. Report No. 81-10, Department of Civil Engineering, University of Missouri, Rolla.
Chien, L. S. & Sun, C. T. [1989]. Parallel processing techniques for finite element analysis of nonlinear large truss structures. *Computers and Structures,* **31,** 1023–9.
Deoppner, T. [1987]. Threads: a system for the support of concurrent programming. Report CS-87-11, Department of Computer Science, Brown University, Providence, RI.
Dijkstra, E. W. [1965]. Cooperating sequential processes. Technical Report EWD-123, Technological University, Eindhoven, The Netherlands.
Dongarra, J. & Duff, I. S. [1992]. Advanced architecture computers. In *Supercomputing in Engineering Analysis,* ed. H. Adeli. Marcel Dekker, New York, pp. 19–62.
Encore [1985]. *Multimax Technical Summary.* Encore Computer Corporation, Marlboro, MA.
Encore [1988]. *Encore Parallel Threads Manual.* Encore Computer Corporation, Marlboro, MA.

Farhat, C. & Wilson, E. [1987]. Modal superposition analysis on concurrent multiprocessors. *Engineering Computations*, **3**, 305–11.
Farhat, C., Felippa, C. & Park, K. [1987a]. Implementation aspects of concurrent finite element computations. In *Parallel Computations and their Impact on Mechanics*, ed. A. Noor. AMD Vol. 86, American Society of Mechanical Engineers, pp. 301–15.
Farhat, C., Wilson, E. & Powell, G. [1987b]. Solution of finite elements systems on concurrent processing computers. *Engineering with Computers*, **2**, 157–65.
Hoare, C. A. R. [1974]. Montiors: an operating system structuring concept. *Communications of the ACM*, **17**, 549–57.
Kamal, O. & Adeli, H. [1990]. Automatic partitioning of frame structures for concurrent processing. *Microcomputers in Civil Engineering*, **5**, 269–83.
Kamal, O. & Adeli, H. [1991]. A substructuring algorithm for analysis and optimization of large structures on parallel machines. In *Mechanics Computing in 1990's and Beyond—Volume One—Computational Mechanics, Fluid Mechanics, and Biomechanics*, ed. H. Adeli & R. L. Sierakowski. American Society of Civil Engineers, New York, pp. 68–72.
Kernighan, B. & Ritchie, D. [1988]. *The C Programming Language*, 2nd edn. Prentice-Hall, Englewood Cliffs, NJ.
Khan, M. R., Wilmert, K. D. & Thornton, W. A. [1979]. An optimality criterion method for large-scale structures. *AIAA Journal*, **17**, 753–61.
Khot, N. S., Berke, L. & Venkayya, V.B. [1979]. Comparison of optimality criteria algorithms for minimum weight design of structures. *AIAA Journal*, **17**, pp. 182–190.
Kochan, S. [1988]. *Programming in C*, rev. edn. Hyden Books, IN.
Leger, P., Wilson, E. L. & Clough, R. W. [1986]. The use of load dependent vectors for dynamic and earthquake analysis. Report UCB/EERC-86/04, Earthquake Engineering Research Center, University of California, Berkeley.
Lo, S. & Phillipe, B. [1986]. The symmetric eigenvalue problem on a multiprocessor. In *Parallel Algorithms and Architectures*, ed. M. Cosnard, P. Quinton, Y. Robert & M. Tchuente. Elsevier Science Publishers, New York, pp. 31–43.
Lou, J. C. & Friedman, B. [1989]. Parallel algorithms for the finite element method. *Mechanics Research Communications*, **16**, 3–18.
Noor, A. K. & Peters, J. M. [1989]. A partitioning strategy for efficient nonlinear finite element dynamic analysis on multiprocessor computers. *Computers and Structures*, **31**, 795–810.
Noor, A. K., Kamel, N. A. & Fulton, R. E. [1978]. Substructuring techniques—Status and projections. *Computers and Structures*, **8**, 621–32.
Peterson, J. & Silberschatz, A. [1985]. *Operating System Concepts*, 2nd edn. Addison-Wesley, Reading, MA.
Saad, Y. [1986]. Gausslian elimination on hypercubes. In *Parallel Algorithms and Architectures*, eds. M. Cosnard, P. Quinton, Y. Robert & M. Tchuente. Elsevier Science Publishers, New York, pp. 5–17.
Sikiotis, E. S. & Saouma, V. E. [1987]. Parallel structural optimization on a computer network. In *Proceedings of the ASCE Conference on Structural engineering, Orlando, FL*.

Svensson, B. [1987]. A substructuring approach to optimum structural design. *Computers and Structures*, **25**, 251–8.
Venkayya, V. B., Khot, N. S. & Reddy, V. S. [1969]. Energy distribution in an optimum structural design. AFFDL-TR-68-156, Wright Patterson Air Force Base, Dayton, OH.
Vishnubhotla, P. & Adeli, H. [1992]. Parallel programming languages and techniques. In *Parallel Processing in Computational Mechanics*, ed. H. Adeli, Marcel Dekker, New York, pp. 21–32.

Index

Active column storage method, 78
Active displacement constraints
  effect on optimal designs, 153,
    154, 155, 156
Algorithms
  concurrent reduction of
    subdomains, 86–7
  concurrent solution of interface
    linear equations, 89–90
  concurrent solution of reduced
    eigenproblem, 151
  evaluation of maximum
    displacement ratio, 120
  evaluation of maximum stress
    ratio, 120
  evaluation of new vector of
    design variables, 121
  generation of Ritz vectors, 149
  parallel matrix operations, 150–1
  parallel optimization of framed
    structures
    under combined loadings,
      145–7
    under static loading, 118–19
  parallel temporal solution, 152
  partitioning
    applications of, 60–8
    final partitioning, 53
    initial partitioning, 45
    intermediate partitioning, 51
Amdahl's second law, 35
Autotasking, 178

Backward substitution, 88, 130
  algorithm for, 89–90
  parallelism in, 91
Banded stiffness matrix, 21
Basic concepts (in parallel
  processing), 3–31
Border nodes
  meaning of term, 37
  see also Interface nodes
Boundary conditions, application
  of, 7, 83, 128–9
Boundary nodes
  meaning of term, 37
  see also Support nodes
Braced frame (90-element)
  concurrent optimization under
    dynamic loading
    overall efficiency of, 170
    overall speed-ups in, 168, 169,
      173
    workload balance for, 170
  concurrent optimization under
    static loading
    overall efficiency of, 139
    overall speed-ups in, 135, 136
    workload balance for, 137
  degrees of freedom for, 136, 169,
    173
  dimensions of, 61
  material properties for, 126, 155
  optimal (minimum weight) design
    of, 126–7, 154–5, 156

Braced frame (90-element)—*contd.*
  partitioning of, 60–1
    4-processor case, 63
    7-processor case, 64
    summary of results, 62
  speed-ups for
    concurrent optimization under dynamic loading, 168, 169, 170, 172, 173
    concurrent optimization under static loading, 135, 136

C programming language, 5
  advantages of, 5
  parallel optimization program in, macro flow chart of, 112
  parallel structural analysis program in, 12–14
  flow chart of, 93
Cheap concurrency, 7, 175
Cholesky decomposition approach, 83
Closed nodes, 37
  see also Internal nodes
Concurrent analysis of structures, 74–110
  applications, 92–107
  computational steps for, 75
  conclusions, 176
  flow chart of parallel program for, 93
Concurrent calculations, percentage time spent on, 16–17, 157
Concurrent optimization of structures
  under dynamic loading, 140–74
    algorithms for, 145–7, 149–52
    applications, 152–71
    conclusions, 177
    flow chart of parallel program for, 144
  under static loading, 111–39
    algorithms for, 117–22
    applications, 122–38
    conclusions, 176
    flow chart of parallel program for, 112

Concurrent processing, 1
  issues involved, 8–12
  see also Parallel processing
Connectivity strategy, 46, 73
Context-switching, 11
  time spent in, 26
Cray Y-MP computer, 2, 177
Critical section problem, 9
  remedies for, 9–10

Displacement (stiffness) method
  steps in structural analysis using, 74, 75
  see also Stiffness (displacement) method
Divide-and-conquer paradigm, 32, 33
Domain partitioning, 32
  example for four-storey three-bay frame, 34, 50, 52, 56, 77
Duhamel integral, 142
Dynamic displacement vectors, calculation of, 152
Dynamic loading
  concurrent optimization under, 140–74
  conclusions, 177
  optimization under, 140–3

Earthquake data, optimal design affected by, 171, 172, 173
Efficiency, 95
Eigenproblem. *See* Reduced eigenproblem...
El Centro earthquake data, 171
  optimal design affected by, 171, 172, 173
Element forces, computation of, 6, 92
  application to 200-bar plane truss, 132–3, 134
  application to 266-element frame, 103
  in concurrent analysis, 92, 103
  in concurrent optimization, 132–3, 134

Elements–thread mapping, 11, 79
 relative speed-up affected by, 27
 speed-up affected by, 23–6
Encore Multimax multiprocessor
   computer, 3–4
 architecture of, 3–4
 operating systems supported, 5
 processor used, 4
 shared memory in, 4, 34, 55
Encore Parallel Threads, 7–8
 sample code for EPT function
   calls, 179
Equilibrium equations, 5
 dynamic loading, 140–1
 static loading, 76, 141
Explicit pivoting procedure, 88

Factorization (of stiffness matrix), 84
 percentage time spent on, 130
Final partitioning, 52–7
 algorithm for, 53
 equilibrium equations resulting, 76
 flow chart for, 54
 purpose of, 52
 speed-ups at end of, 98, 99, 101, 102, 104–6
Fine-grained parallelism
 effects of, 100, 169, 176
 minimization of, 90, 101
Finite-element analysis, steps
   involved, 6, 32–3
Flow charts
 final partitioning, 54
 initial partitioning, 46
 intermediate partitioning, 51
 parallel optimization program
   dynamic-loading case, 144
   static-loading case, 112
 parallel structural analysis
   program, 93
 partitioning algorithm, 41
Forces and stresses, computation of, 6, 92
 application to 200-bar plane truss, 132–3, 134

Forces and stresses, computation of—*contd.*
 application to 266-element frame, 103
 in concurrent analysis, 92, 103
 in concurrent optimization, 132–3, 134
Forty-storey frame, 94
 see also Plane frame (760-element. . .
Forward substitution, 88, 130
 algorithm for, 89
 parallelism in, 91
Four-storey three-bay frame
 characteristic arrays for
   at end of initial partitioning, 47–8
   at end of pre-partitioning, 43–4
 domain partitioning of, 34, 49, 52, 56, 77
 numbering scheme for, 36
 partitioned structure
   at end of final partitioning, 56
   at end of initial partitioning, 49
   at end of intermediate
     partitioning, 52
 partitioning of, 34, 37, 39–40, 43–4, 47–8, 50, 52, 56, 58–9
 equilibrium equations resulting, 76
 renumbering scheme for
   at end of final partitioning, 59, 77
   at end of initial partitioning, 50
   at end of post-partitioning, 58
 stiffness matrix for
   at end of final partitioning
     stage, 59, 77
   at end of initial partitioning, 50
Framed structures
 concurrent analysis of, 74–110
 concurrent optimization of, under
   dynamic loading, 140–74
 concurrent optimization under
   static loading, 111–39
 partitioning of, 32–73
Future research proposed, 177–8

Gaussian elimination procedure, 83
Generalized load vector,
 composition of, 6
Geodesic dome space truss (132-
 element)
 concurrent analysis of
  overall efficiency of, 105, 106,
   109
  overall speed-up for, 104, 105,
   106
  workload balance for, 105, 106,
   108
 concurrent optimization under
  dynamic loading
  overall efficiency of, 170
  overall speed-ups in, 168, 169,
   173
  workload balance for, 170
 concurrent optimization under
  static loading
  overall efficiency of, 139
  overall speed-ups in, 135, 136
  workload balance for, 137
 degrees of freedom of, 105, 106,
  136
 dimensions of, 65
 material properties for, 125,
  155–6
 numbering of elements and nodes
  in, 65
 optimal (minimum weight) design
  of, 124–5, 155–6, 157
 partitioning of, 61, 63
  3-processor case, 66
  4-processor case, 66
  summary of results, 65
 speed-ups for
  concurrent analysis, 104, 105,
   106
  concurrent optimization under
   dynamic loading, 168, 169,
   173
  concurrent optimization under
   static loading, 135, 136
Geodesic dome structure (156-
 element)
 degrees of freedom of, 14
 numbering of elements and nodes
  in, 15

Geodesic dome structure (156-
 element)—*contd.*
 PASTRANC performance for, 29
 relative speed-up for
  effect of elements–thread
   mapping, 27
  effect of synchronization
   mechanisms, 28
  effect of thread-creation
   overhead time, 28
 speed-up for
  effect of elements–thread
   mapping, 24, 25
  effect of synchronization
   mechanisms, 22
  effect of threads-creation
   overhead time, 18, 19
  time spent (as percentage) on
   concurrent calculations, 17

Initial partitioning, 44–9
 algorithm for, 45
 arrays set up at end of, 44–5
 flow chart for, 46
 purpose of, 44
 speed-ups at end of, 99, 102, 103
Inner product of vectors, algorithm
 for, 150
Intel hypercube computer, 2, 34, 56
Interface displacements
 internal-node displacements
  expressed in terms of, 85
 see also Non-interface
  displacements
Interface linear equations
 concurrent solution of, 88–91,
  130, 132, 173
 algorithm for, 89–90
 application for 200-bar plane
  truss, 132
 application for 266-element
  frame, 100–1
 parallelism in, 91
 generation of, 85
Interface nodes
 meaning of term, 37
 number related to locations
  requiring synchronization,
  81

Interface nodes—*contd.*
  number related to number of
      processors
    200-bar plane truss, 131
    266-element frame, 98
    798-element frame, 159
  number related to number of
      semaphores, 81
  reduction in number of, 41, 46
  switched to internal nodes, 49
Intermediate partitioning, 49,
    50–2
  algorithm for, 51
  flow chart for, 51
  purpose of, 49, 51
Internal nodes
  balancing among subdomains,
      54–5
  displacements expressed in terms
      of interface displacements,
      85
  interface nodes switched to, 49
  meaning of term, 37
  number related to number of
      processors
    200-bar plane truss, 131
    266-element frame, 98
    798-element frame, 159
  numbering of, 37, 38

Kuhn–Tucker conditions, 114, 116

Lagrangian multipliers, 114
  formula for estimation of, 116
Load vector, assembly of, 7, 82–3

Macro flow charts
  optimization program
    dynamic-loading case, 144
    static-loading case, 112
  partitioning algorithm, 41
Macrotasking, 178
Mapping, 10–12
  elements onto threads, 11
  subdomains onto processors, 36,
      37, 38

Mapping strategies, 11–12
  in concurrent analysis, 76
  in domain partitioning, 36–7
  in PASTRANC program, 14
  relative speed-up affected by, 27
  speed-up affected by, 23–6
Mass matrix. *See* Structure mass
    matrix
Matrix operations, 88, 89–90, 130
  parallel matrix operations, 150–1
  see also Backward. . . ; Forward
      substitution; Inner
      product; Multiplication;
      Normalization;
      Orthogonalization; Upper
      triangularization
Microtasking, 178
MIPS, 1
Monitors, 10
  cases when used, 10, 31
  relative speed-up affected by, 28,
      31
  speed-up affected by, 21–3, 31
Multiplication of matrices,
    algorithm for, 150
Multiprocessor computers, 2
  see also Cray Y-MP. . . ; Encore
      Multimax. . . ; Intel
      hypercube. . .
Multitasking, 178

Nanobus, 4
Non-interface displacements,
    computation of, 85, 92
  speed-ups for 266-element frame,
      102
Normalization of vector, algorithm
    for, 151

Open nodes
  meaning of term, 37
  see also Interface nodes
Optimal design examples, 123–8,
    153–7
Optimality criterion approach,
    113–16

Optimization
 concurrent optimization
  under dynamic loading, 140–74
  under static loading, 111–39
 dynamic loading case, 140–3
 macro flow chart for parallel C
  program, 112
 see also Concurrent optimization
Orthogonalization of vector,
 algorithm for, 150
Overhead time
 context-switching, speed-up
  affected by, 26, 31
 invocation of synchronization
  mechanisms, 21
 speed-up affected by, 21–3, 31
 in static condensation step, 99
 thread-creation, 18
 relative speed-up affected by,
  28
 speed-up affected by, 18–20

Parallel matrix operations, 150–1
Parallel processing
 advantages of, 1
 basic concepts, 3–31
 application of, 14–30
 issues involved, 8–12
 multiprocessor computers used, 2
Parallel structural optimization
 algorithms for, 117–22
 applications, 122–38
Parallelization of code
 computational steps not
  amenable, 34–5
 in PASTRANC, 12, 13, 14
Partitioning, 32–73
 basic concepts and definitions,
  36–42
 final partitioning, 52–7
 initial partitioning, 44–9
 intermediate partitioning, 49,
  50–2
 meaning of term, 32
 post-partitioning, 57–8, 60
 pre-partitioning, 42–4
 stages in, 41

Partitioning—contd.
 see also Final...; Initial...;
  Intermediate...; Post-
  ...; Pre-partitioning
Partitioning algorithm
 applications, 60–8
  geodesic dome space truss, 61,
   63
  90-element braced frame, 60–1
  200-bar plane truss, 63–4, 66
  266-element plane frame, 66–8
 effectiveness demonstrated, 99,
  100, 102, 103–5
 features required, 37–9, 41, 176
 final-partitioning algorithm, 53
 initial-partitioning algorithm, 45
 intermediate-partitioning
  algorithm, 51
 macro flow chart of, 41
 pre-processing before use, 42–4, 72
PASTRANC (PArallel STRuctural
 ANalysis in C) program,
 12–14
 applications, 14–30
 mapping strategies used, 14
 options preferred, 30
 overall time performance, 28–30
 parallelization of portions of
  code, 12, 13, 14
 percentage time spent on
  concurrent calculations,
  16–17
 speed-up versus mapping, 23–6
 synchronization mechanisms
  used, 14
Pivoting procedure, 88
Plane frame (28-element). See Four-
 storey three-bay frame
Plane frame (266-element irregular
 frame), 70
 concurrent analysis of, 95–104
 overall efficiency of, 105, 106,
  108
 overall speed-up for, 104, 105,
  106
 workload balance for, 105, 106,
  107
 degrees of freedom of, 105, 106

Plane frame (266-element irregular frame)—*contd.*
 dimensions of, 70
 interface linear equations solved for, speed-up for, 101
 non-interface displacements computed for, 102
 numbers of interface and internal nodes for, 98
 partitioning of, 66–8
  6-processor case, 70
  10-processor case, 72
  summary of results, 71
 speed-ups for
  concurrent analysis, 104, 105, 106
  interface linear equations solution, 101
  static condensation, 98, 99–100
  stiffness-matrix assembly, 95–7
 static condensation step for, 97–100
  speed-ups for, 98, 99–100
  time analysis for, 97
 stiffness matrix assembled for, 95
  speed-ups for, 95–7
Plane frame (760-element 40-storey frame), 94
 concurrent analysis of
  overall efficiency of, 105, 106, 109
  overall speed-up for, 104, 105, 106
  workload balance for, 105, 106, 108
  degrees of freedom of, 105, 106
Plane frame (798-element 60-storey irregular frame), 123
 concurrent optimization under dynamic loading, 158–71
  overall efficiency of, 169, 170
  overall speed-ups in, 168, 169, 172, 173
  workload balance for, 160–2, 166–8, 170
 concurrent optimization under static loading
  overall efficiency of, 138

Plane frame (798-element 60-storey irregular frame)—*contd.*
 concurrent optimization under static loading—*contd.*
  overall speed-ups in, 135, 136
  workload balance for, 137
 degrees of freedom of, 136, 169, 173
 dimensions of, 153
 material properties for, 127, 153
 numbers of interface and internal nodes for, 159
 optimal (minimum weight) design of, 127, 128, 153–4
 speed-ups for
  concurrent analysis, 105, 106
  concurrent optimization under dynamic loading, 168, 169, 172, 173
  concurrent optimization under static loading, 135, 136
 speed-ups for eigenvector computation, 164
 gradient-evaluation step, 167
 maximum-response evaluation, 166
 reduced-eigenproblem solution, 163
 Ritz-vector evaluation, 161
 static analysis, 159
 structure mass matrix assembly, 160
 temporal-solution step, 164
Plane truss (200-bar truss)
 boundary-conditions application for, speed-up for, 129
 concurrent analysis of
  overall efficiency of, 105, 106, 109
  overall speed-up for, 104, 105, 106, 160
  workload balance for, 105, 106, 108
 concurrent optimization under dynamic loading
  overall efficiency of, 170
  overall speed-ups in, 168, 169, 172, 173

Plane truss (200-bar truss)—*contd.*
  concurrent optimization under dynamic loading—*contd.*
    workload balance for, 170
  concurrent optimization under static loading, 128–34
    overall efficiency of, 138
    overall speed-ups in, 134, 136
    workload balance for, 137
  degrees of freedom of, 105, 106, 169, 173
  dimensions of, 67, 123
  load-vector assembly for, speed-up for, 129
  material properties for, 123, 154
  numbers of interface and internal nodes for, 159
  optimal (minimum weight) design of, 123–4, 154, 155
  partitioning of, 63–4, 66
    4-processor case, 69
    8-processor case, 69
    summary of results, 68
  Ritz-vector evaluation, speed-ups for, 162
  speed-ups for
    boundary-conditions application, 129
    concurrent analysis, 104, 105, 106, 160
    concurrent optimization under dynamic loading, 168, 169, 172, 173
    concurrent optimization under static loading, 134, 136
    load-vector assembly, 129
    Ritz-vector evaluation, 162
    static-condensation step, 131
    stiffness-matrix assembly, 129
  static condensation step for
    percentage time spent on each stage, 130
    speed-ups for, 131
  stiffness-matrix assembly for, speed-up for, 129
Post-partitioning, 57–8, 60
  purpose of, 57
  vectors defined, 57

Pre-partitioning, 42–4, 72
  arrays set up at end of, 42
Programming language, 5

Racing condition, 8–9, 79
  remedies for
    comparison of strategies in static condensation step, 98, 99
    element-renumbering suggested, 79
    extra storage locations used, 81, 88, 95, 96, 117, 148, 176
    synchronization used, 8, 9, 20–1, 79, 81, 88, 95, 117, 176
Reduced eigenproblem, solution of, 148–9, 161–3
  algorithm for, 151
  speed-ups for 798-element frame, 163
Relative speed-up
  effect of mapping strategies, 27
  effect of synchronization mechanisms, 28
  effect of thread-creation overhead time, 28
  meaning of term, 15
Ritz vectors
  evaluation of, 161
  speed-ups for, 161, 162
  generation of, 148, 161
  algorithm for, 149

Seismic loading, 140
  see also Dynamic loading; Earthquake data
Semaphores, 9–10
  cases when used, 10, 20–1, 31
  relative speed-up affected by, 28, 31
  sample code for EPT function call, 179
  speed-up affected by, 20–3, 31

Shared-memory multiprocessor
  computers
  advantages of, 34, 55
  memory management in, 4, 175
  parallel matrix operations on,
    150–1
  Ritz vectors generated for use on,
    149
  see also Encore Multimax...
Sixty-storey frame, 123
  see also Plane frame (798-
    element...
Skyline storage method, 78–9, 80
Space truss (72-bar)
  degrees of freedom of, 14
  numbering of elements and nodes
    in, 16
  PASTRANC performance for, 29
  speed-up for
    effect of elements–thread
      mapping, 24, 25
    effect of synchronization
      mechanisms, 22
    effect of threads-creation
      overhead time, 18, 19
    time spent (as percentage) on
      concurrent calculations, 17
Space truss see also Geodesic
  dome...
Speed-ups
  active displacement constraints
    gradients evaluation, 167,
    168
  boundary-conditions application,
    129
  concurrent-analysis overall,
    103–6, 159–60
  concurrent-optimization overall
    under dynamic loading,
      168–70, 172–3
    under static loading, 133–6
  effect of mapping strategies, 23–6
  effect of semaphores versus
    monitors, 20–3
  effect of threads-creation
    overhead time, 18–20
  effect of unparallelizable code, 35
  eigenvectors computation, 164

Speed-ups—*contd.*
  element-forces/stresses
    computation, 103
  interface linear equations
    solution, 100, 101
  load-vector assembly, 129
  mass-matrix assembly, 160
  maximum-response evaluation,
    166
  meaning of term, 15, 95
  non-interface displacements
    computation, 102
  reduced-eigenproblem solution,
    163
  Ritz-vectors evaluation, 161, 162
  static condensation, 98, 99, 131
  static (structural) analysis overall,
    104, 105, 106, 159, 160
  stiffness-matrix assembly, 95–6,
    129
  temporal solution, 164
Static condensation, 83–8
  application
    200-bar plane truss, 129–30, 131
    266-element frame, 97–100
Static loading
  concurrent optimization under,
    111–39
  conclusions for, 176
Static (structural) analysis, 74–111
  applications, 92–107
  computational steps listed, 6,
    32–3, 75
  in concurrent optimization
    under dynamic loading, 144,
      158
    under static loading, 112, 117
  flowchart of parallel C program
    for, 93
  speed-ups for, 104, 105, 106, 159,
    160
Stiffness (displacement) method,
  structural analysis using,
  steps of, 74, 75
Stiffness matrix. See Structure
  stiffness matrix
Stresses. See Forces and stresses
Stress-ratio recursion formula, 116

Structural analysis
  computational steps involved, 6, 32–3
  parallelization of, 33–4
  see also Boundary conditions. . . ; Element forces. . . ; Interface linear equations. . . ; Load vector. . . ; Non-interface displacement. . . ; Static condensation. . . ; Stress vectors. . . ; Structure stiffness matrix. . .
  linear problem summarized, 5–7
Structure mass matrix
  assembling of, 143, 148
  algorithm for, 147
  speed-ups for 798-element frame, 158, 160
Structure stiffness matrix
  after partitioning, 37, 39
  assembling of, 6, 78–82
  algorithm for, 82
  application for 266-element frame, 95–7
  speed-ups for 266-element frame, 95–6
  banded form, 21
  compact form, 78, 79
  decoupled form, 78
  four-storey three-bay frame example, 50, 59, 77, 78
  storage methods for, 78–9
  typical form, 20–1
Subdomain, meaning of term, 42
Subdomaining. See Partitioning
Substructure, meaning of term, 42
Substructuring. . . See Partitioning. . .
Substructuring algorithm
  applications, 60–8
  effectiveness demonstrated, 99, 100, 102, 103–5
  features required, 37–9, 41, 176
Support nodes
  meaning of term, 37
  numbering of, 37, 38

Synchronization (of processors), 8, 9–10
  mechanisms used, 9–10
  in PASTRANC program, 14
  relative speed-up affected by, 28
  sample code for EPT function call, 179
  speed-up affected by, 20–3, 79, 175
  see also Monitors; Semaphores

Temporal solution, 149, 151
  algorithm for, 152
Threads, 7–8, 175
  concept first developed, 7
  creation of, sample code for EPT function call, 179
  meaning of term, 7
  overhead time required for creation of, 7, 18
  relative speed-up affected by, 28
  speed-up affected by, 18–20, 26, 175
Tridiagonal matrix, construction of, 148, 149

Unparallelizable code, effect on speed-up, 35
Upper triangularization, 88, 130
  algorithm for, 89
  parallelism in, 91

Vector machine, 177–178
Vector processing, 178

Wilson–Ritz algorithm, parallelization of, 148
Wilson-Θ direct integration method, 142, 151

Workload balance
in concurrent analysis, 88, 91, 102, 105, 106–7, 108, 159, 160
in concurrent optimization
under dynamic loading, 160–2, 166–8, 170
under static loading, 129, 131,

Workload balance—*contd.*
in concurrent optimization—*contd.*
134, 136, 137
meaning of term, 95
need for, 41
partitioning algorithm affecting, 46, 48, 72